Mapping Workbook for
World Regional Geography
Fourth Edition

Jennifer Rogalsky
State University of New York, College at Geneseo

Helen Ruth Aspaas
Virginia Commonwealth University

W. H. Freeman and Company
New York

Media and Supplements Editor: Deepa Chungi
Project Editor: Jodi Isman
Production Manager: Julia DeRosa
Map Designer: Will Fontanez
Marketing Director: Scott Guile

ISBN-13: 978-1-4292-0499-6
ISBN-10: 1-4292-0499-0

Printed in the United States of America

First printing

W. H. Freeman and Company
41 Madison Avenue
New York, NY 10010
Houndmills, Basingstoke
R621 6XS, England
www.whfreeman.com

Contents

CHAPTER ONE

Geography: An Exploration of Connections

MAPPING EXERCISES

The following are three mapping exercises to improve your knowledge of the location of places, underscore why they are important, and clarify how they relate to each other. Some questions will ask you to locate places, compare maps, or fill in data; others will test your understanding of *why* you were asked to map the features that you did. Use the blank outline maps at the end of the chapter to complete these exercises. Additional blank outline maps can be found on the textbook's Web site: www.whfreeman.com/pulsipher4e.

1. Is there a digital divide between North America and Central America?

The digital divide refers to the division between people who use computers and have access to the Internet and those who do not.

- Use the blank maps for North America and Central America to complete this mapping exercise.
- Visit the CIA World Factbook at www.cia.gov/cia/publications/factbook.
- Obtain information for the GDP per capita and the number of Internet users for each of the following countries: Canada, the United States, Mexico, Belize, Guatemala, Honduras, Nicaragua, El Salvador, Costa Rica, and Panama.
- Choose two different symbols to map the GDP per capita and the number of Internet users. Shade each country with a shade of gray (darkest shade of gray representing the highest GDP per capita). Use graduated circles to portray the number of Internet users.

Questions

a. What is the relationship between the GDP per capita for each country and its number of Internet users? Explain the relationship.

b. Based on your maps, can you say that the digital divide exists? If it does, is there a clear division between those countries that are rich and have access to the Internet and those countries that are poor and have minimum access to the Internet? If not, try to explain why this is the case.

c. For poor countries, how would stepping across the digital divide and gaining more Internet access help improve living conditions for the poor?

2. Arable land, agriculture, and soil degradation

Soil erosion appears to be an environmental concern in many countries, but it may be more serious in those countries where the economy has a high dependency on agriculture. All the countries in South Asia are facing soil degradation issues.

- Visit the CIA World Factbook at www.cia.gov/cia/publications/factbook.

- Collect information on percent of arable land, percent of population who earn their living from agriculture, and the GDP per capita for the countries of South Asia: India, Pakistan, Bangladesh, Afghanistan, Nepal, Bhutan, and Sri Lanka.
- Select appropriate symbols to map these three sets of information on the blank map of South Asia. You may want to use shading for one variable, cross-hatching for another variable, and a symbol of your choice for the third variable.

Questions

a. From your map, identify at least three challenges that farmers face if they want to slow the rate of soil degradation occurring on their farms.
b. Identify the country that may have the most serious challenges in arresting soil degradation and explain why you selected that particular country.

3. Climate classifications

Climate, wind, and weather are largely the result of complex patterns of air temperature and pressure, ocean temperature and currents, and the location of certain landforms.

Questions

a. Consider the climate of where you currently live. List the general climate characteristics including temperature, precipitation, and seasonality characteristics.
b. Next look at Figure 1.22 (precipitation), and Figure 1.25 (climate regions). Write down the description or classification from each figure that corresponds to your area.
c. If the description or classification matches how you described your area, attempt to explain your climate characteristics. If the description or classification does not match yours well, explain why it does not match. Consider the scale of the maps in Figures 1.22 and 1.25, as well as any mountain ranges or bodies of water that may affect climate.

NORTH AMERICA
0 Miles 500
0 Kilometers 800

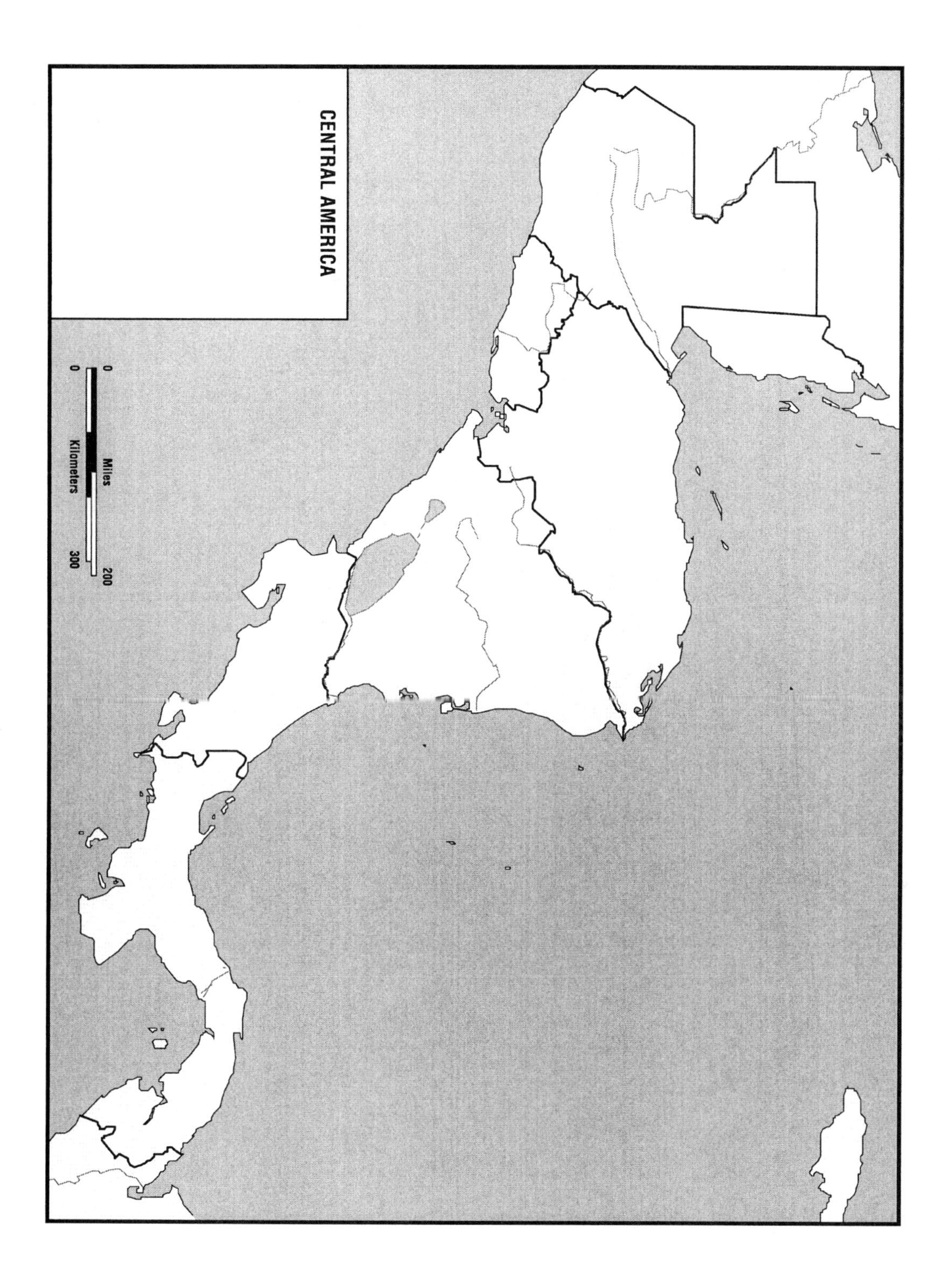
CENTRAL AMERICA
Miles
0
200
Kilometers
0
300

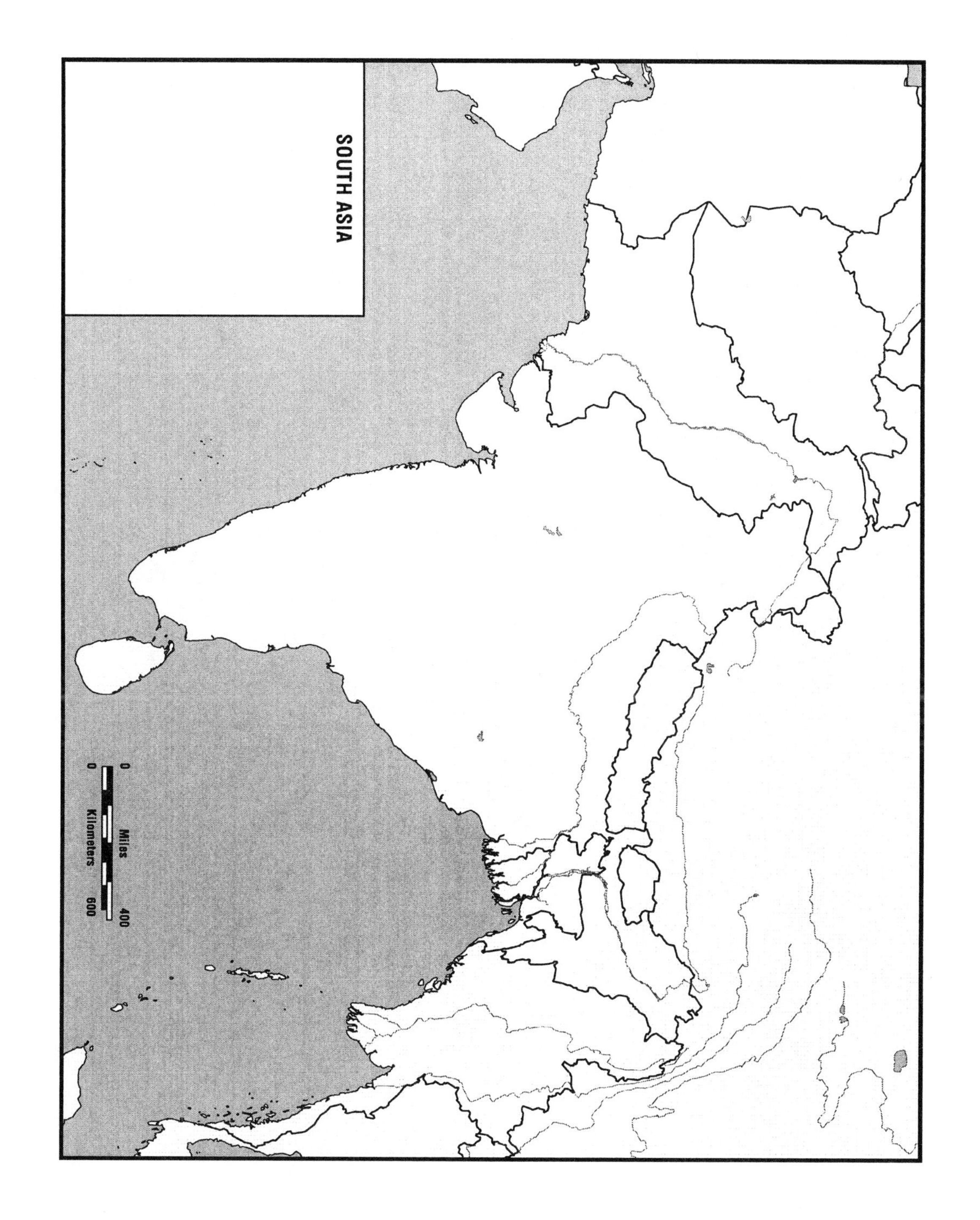
SOUTH ASIA
Miles
0
400
Kilometers
0
600

CHAPTER TWO
North America

IMPORTANT PLACES

The following places are featured in the chapter. Make sure you can locate all of them on a map. A blank outline map of the region is provided at the end of this chapter; additional maps can be found on the textbook's Web site: www.whfreeman.com/pulsipher4e. Also, to prepare for exams, write a few important facts about each place in the space provided.

Physical Features

1. Appalachian Mountains

2. Atlantic Ocean

3. Bering land bridge

4. Canadian Shield

5. Catskill-Delaware watershed

6. Colorado River

7. Columbia River

8. Grand Banks

9. Great Basin

10. Great Lakes

11. Great Plains

12. Gulf of Mexico

13. Hudson Bay

14. James Bay

15. Lake Michigan

16. Mississippi River Valley and Delta

17. Missouri River

18. Ogallala Aquifer

19. Ohio River

20. Pacific Ocean

21. Peace River

22. Rocky Mountains

23. Snake River

24. St. Lawrence River

25. Utah Valley

Regions/Countries/States/Provinces

26. Alabama

27. Alaska

28. Alberta

29. Appalachia

30. Arizona

31. Arkansas

32. Atlantic Provinces

33. British Columbia

34. California

35. Colorado

36. Connecticut

37. Continental Interior

38. Delaware

39. Florida

40. Georgia

41. Great Plains Breadbasket

42. Idaho

43. Illinois

44. Indiana

45. Iowa

46. Kansas

47. Kentucky

48. Louisiana

49. Maine

50. Manitoba

51. Maritime Provinces

52. Maryland

53. Massachusetts

54. Michigan

55. Midwest (Middle West)

56. Minnesota

57. Mississippi

58. Missouri

59. Montana

60. Nebraska

61. Nevada

62. New Brunswick

63. New England

64. Newfoundland

65. New Hampshire

66. New Jersey

67. New Mexico

68. New York

69. North Carolina

70. North Dakota

71. Northeast

72. Northwest Territories

73. Nova Scotia

74. Nunavut

75. Ohio

76. Oklahoma

77. Old Economic Core

78. Ontario

79. Oregon

80. Pacific Northwest

81. Pennsylvania

82. Prairie Provinces

83. Prince Edward Island

84. Quebec

85. Rhode Island

86. Saskatchewan

87. South Carolina

88. South Dakota

89. Southeast (The American South)

90. Southwest

91. Tennessee

92. Texas

93. Utah

94. Vermont

95. Virginia

96. Yukon Territories

97. Washington

98. The (American) West

100. West Coast

101. West Virginia

102. Wisconsin

103. Wyoming

Cities/Urban Areas

104. Atlanta, GA

105. Austin, TX

106. Baltimore, MD

107. Bentonville, AR

108. Birmingham, AL

109. Boston, MA

110. Calgary, Alberta

111. Chicago, IL

112. Cincinnati, OH

113. Cleveland, OH

114. Dallas, TX

115. Denver, CO

116. Detroit, MI

117. El Paso, TX

118. Eugene, OR

119. Georgetown, KY

120. Hackensack, NJ

121. Hartford, CT

122. Indianapolis, IN

123. Jamestown, VA

124. Kansas City, MO

125. Laramie, WY

126. Lewiston, ME

127. Laredo, TX

128. Los Angeles, CA

129. Lowell, MA

130. Megalopolis

131. Miami, FL

132. Minneapolis, MN

133. Montreal, Quebec

134. Nashville, TN

135. New Orleans, LA

136. New York, NY

137. Nogales, AZ

138. North Lawndale, IL

139. Philadelphia, PA

140. Phoenix, AZ

141. Pittsburgh, PA

142. Pittsford, NY

143. Portland, OR

144. Providence, RI

145. Québec City, Québec

146. Raleigh-Durham, NC

147. Rochester, NY

148. Salt Lake City, UT

149. San Diego, CA

150. San Francisco, CA

151. Seattle, WA

152. St. Louis, MO

153. Storm Lake, IA

154. Terrebone Parish, LA

155. Toronto, Ontario

156. Valdez, AL

157. Vancouver, British Columbia

158. Washington, D.C.

MAPPING EXERCISES

The following are three mapping exercises to improve your knowledge of the location of places, underscore why they are important, and clarify how they relate to each other. Some questions will ask you to locate places, compare maps, or fill in data; others will test your understanding of *why* you were asked to map the features that you did. Use the blank outline maps at the end of the chapter to complete these exercises. Additional blank outline maps can be found on the textbook's Web site: www.whfreeman.com/pulsipher4e.

1. Regional climate, population density, and agriculture practices

Agriculture in North America is highly productive, but usually in areas with favorable climates and abundant natural resources. Refer to the climate map (Figure 2.6), the population distribution map (Figure 2.12), and the agriculture map (Figure 2.18) to answer the following questions.

Questions

a. What are two possible human and two physical geography explanations for the location of the *mixed farming* activities?
b. What are two possible human and physical geography explanations for the location of the *range livestock* activities?
c. What are two possible human and physical geography explanations for the location of *corn belt, cash grain*, and *livestock* activities?

2. Mobility and aging in developed societies

For a multitude of reasons, the populations of this region are highly mobile. In some cases, mobility can increase as one ages, especially for those eager to enjoy the amenities of warm climates, say in Arizona or Florida. Use the population by region map (Figure 2.13) and the changing distribution of elderly map (Figure 2.37) to answer the following questions.

Questions

a. Which regions of the United States have experienced the largest increase in population between the years of 1900 and 2000? Provide at least three reasons to explain this noteworthy redistribution of the U.S. population.

b. It was noted previously that Arizona and Florida attract many senior citizens who move to these states to enjoy retirement. How, then, do we explain the growing population of elderly in the Great Plains Breadbasket, the Rust Belt, and New England?

3. Environmental issues in North America

North America's environmental issues vary across the region. Some issues are highly localized while others affect large populations. Refer to the air and water pollution map (Figure 2.38), the population density map (Figure 2.12), the agriculture map (Figure 2.18), and the human impact map (Figure 2.39) to answer the following questions.

Questions

a. Which environmental issues are likely to be associated with high population densities? Why is this the case?

b. Which environmental issues are not likely to be associated with high population densities? Why is this the case?

NORTH AMERICA
0 Miles 500
0 Kilometers 800

CHAPTER THREE
Middle and South America

IMPORTANT PLACES

The following places are featured in the chapter. Make sure you can locate all of them on a map. A blank outline map of the region is provided at the end of this chapter; additional maps can be found on the textbook's Web site: www.whfreeman.com/pulsipher4e. Also, to prepare for exams, write a few important facts about each place in the space provided.

Physical Features

1. Amazon Basin

2. Amazon River

3. Andes Mountains

4. Atacama Desert

5. Baja California

6. Brazilian Highlands

7. Caribbean Sea

8. Falkland Islands

9. Galapagos Islands

10. Greater Antilles

11. Guiana Highlands

12. Gulf of Mexico

13. Hispaniola

14. Lake Titicaca

15. Lesser Antilles

16. Orinoco River

17. Pampas

18. Parana River

19. Patagonia

20. Rio Grande River

21. Sierra Madre Occidental

22. Sierra Madre Oriental

23. Tierra del Fuego

24. Yucatan Peninsula

Regions/Countries/States/Provinces

25. Antigua and Barbuda

26. Argentina

27. Bahamas

28. Barbados

29. Belize

30. Bolivia

31. Brazil

32. Chile

33. Colombia

34. Costa Rica

35. Cuba

36. Dominica

37. Dominican Republic

38. Ecuador

39. El Salvador

40. French Guiana

41. Grenada

42. Guadaloupe

43. Guatemala

44. Guyana

45. Haiti

46. Honduras

47. Jamaica

48. Martinique

49. Mexico

50. Montserrat

51. Netherlands Antilles

52. Nicaragua

53. Panama

54. Paraguay

55. Peru

56. Puerto Rico

57. St. Kitts and Nevis

58. St. Lucia

59. St. Vincent and the Grenadines

60. Suriname

61. Trinidad and Tobago

62. Uruguay

63. Venezuela

Cities/Urban Areas

64. Asunción

65. Belmopan

66. Bogotá

67. Brasília

68. Buenos Aires

69. Cancun

70. Caracas

71. Cayenne

72. Chiapas

73. Curitiba

74. Fortaleza

75. Georgetown

76. Guatemala City

77. Havana

78. Kingston

79. La Paz

80. Lima

81. Managua

82. Mexico City

83. Montevideo

84. Panama City

85. Paramaribo

86. Port-au-Prince

87. Quito

88. Rio de Janeiro

89. San José

90. San Juan

91. San Salvador

92. Santiago

93. Santo Domingo

94. Sao Paulo

95. Sucre

96. Tegucigalpa

MAPPING EXERCISES

The following are three mapping exercises to improve your knowledge of the location of places, underscore why they are important, and clarify how they relate to each other. Some questions will ask you to locate places, compare maps, or fill in data; others will test your understanding of *why* you were asked to map the features that you did. Use the blank outline maps at the end of the chapter to complete these exercises. Additional blank outline maps can be found on the textbook's Web site: www.whfreeman.com/pulsipher4e.

1. Landforms, the environment, and population density

The region of Middle and South America extends south from the midlatitudes of the Northern Hemisphere, across the equator, nearly to Antarctica. Within this vast expanse, there is a wide variety of highland and lowland landforms; some landforms are more conducive to human settlement than others.

- Draw the general outline and label the following landforms from the regional map of Middle and South America (Figure 3.1): Sierra Madre Occidental, Sierra Madre Oriental, Guiana Highlands, Andes Mountains, Brazilian Highlands, Atacama Desert, and Patagonia.
- Also, trace the Amazon River with a thick blue line, and draw the outline of the Amazon Basin with a lighter blue line.
- Using the map of population density (Figure 3.12), shade the areas (in red) that have over 100 people per square kilometer (over 261 people per square mile).

Questions

a. Analyzing your map, describe the general pattern of population distribution.
b. Explain why some highland areas have high population concentrations. Explain why some highland areas have low population concentrations.
c. Explain why some lowland areas have high population concentrations. Explain why some lowland areas have low population concentrations.

2. Population growth, female literacy, and well-being

Overall, investment in education is not sufficient and health care is generally poor in the region. Populations in most countries are still growing rapidly, which can have dramatic effects on future quality of life.

- Shade the 10 countries featured in Figure 3.13 according to their rate of natural increase (1975-2003): yellow (0.0-1.0); orange (1.1-2.0); and red (2.1 and higher).
- Using the table of human well-being rankings (Table 3.3), write in the female literacy rate for all countries in the region.
- Use a graduated symbol (e.g., small to large circles) to map GDP per capita for all countries in the region. Use the following categories: $0-5,000; $5,001-10,000; and $10,001 and higher (Table 3.3).

Questions

a. Compare these population growth rates, female literacy rates, and GDP per capita to those for the United States. What does this tell you about the growth and quality of life in this region?
b. Analyzing the map, do you see any easily explainable patterns in the distribution of literacy? Why?
c. Draw some conclusions about the relationship between population growth and human well-being (using literacy rate and GDP as indicators of human well-being)? Are there any anomalies? If so, explain them.
d. How do you predict rapid population growth will affect this region in terms of GDP and female literacy?

3. Economy and well-being of Middle and South America

GDP per capita masks the very wide disparity of wealth in the region. Some HDI rankings are higher partly because education is somewhat more available across gender and class. GEM is low overall, but is higher in some countries because their governments support education and equal opportunity for women.

- Using the table of human well-being rankings (Table 3.3), shade countries from light blue to dark blue based on these categories of GDP per capita: $0-5,000; $5,001-10,000; and $10,001 and higher.
- Next draw hatch (///) patterns (most dense hatch for the highest ranking and widest hatch for the lowest ranking) over the shading based on these categories of HDI: 0-59; 60-119; and 120 and higher. Remember, high numbers are actually low rankings.
- Use a graduated symbol (e.g., from large to small squares: use larger squares for higher rankings) to map the GEM, using the following three categories: 0-27; 28-55; and 56 and higher. Again, remember that high numbers are actually low rankings.

Questions

a. Briefly describe the general relationship you would expect for the three variables you mapped.

b. Do you see any discrepancies between GDP and HDI (e.g., high GDP with low [high number] HDI or low GDP with high HDI)? Explain why these discrepancies may exist *in general.*

c. One might assume that a country with high GDP per capita would also have high (low number) HDI and GEM rankings. Which three countries stand out the most (i.e., have the most discrepancies)? Thoroughly explain the discrepancies for these specific countries based on history or current conditions.

MIDDLE & S. AMERICA
0 Miles 1000
0 Kilometers 1600

MIDDLE & S. AMERICA
0 Miles 1000
0 Kilometers 1600

MIDDLE & S. AMERICA
0
Miles
1000
0
Kilometers
1600

MIDDLE & S. AMERICA
0
Miles
1000
0
Kilometers
1600

CHAPTER FOUR
Europe

IMPORTANT PLACES

The following places are featured in the chapter. Make sure you can locate all of them on a map. A blank outline map of the region is provided at the end of this chapter; additional maps can be found on the textbook's Web site: www.whfreeman.com/pulsipher4e. Also, to prepare for exams, write a few important facts about each place in the space provided.

Physical Features

1. Adriatic Sea

2. Alps

3. Arctic Ocean

4. Atlantic Ocean

5. Baltic Sea

6. Black Sea

7. Danube River

8. Elbe River

9. English Channel

10. Faroe Islands

11. French Riviera

12. Iberian Peninsula

13. Mediterranean Sea

14. Morava River

15. North European Plain

16. North Sea

17. Pyrenees

18. Rhine River and Delta

19. Sardinia

20. Sicily

21. Straits of Gibraltar

22. Svalbard

Regions/Countries/States/Provinces

23. Albania

24. Austria

25. Balkans

26. Baltic States

27. Basque country

28. Belgium

29. Benelux

30. Bosnia-Herzegovina

31. British Isles

32. Bulgaria

33. Catalonia

34. Central Europe

35. Champagne

36. Croatia

37. Cyprus

38. Czech Republic

39. Denmark

40. England

41. Estonia

42. Finland

43. France

44. Galicia

45. Germany

46. Great Britain (Britain)

47. Greece

48. Greenland

49. Hungary

50. Iceland

51. Ireland (Republic of)

52. Italy

53. Kaliningrad

54. Kosovo

55. Latvia

56. Lithuania

57. Luxembourg

58. Macedonia

59. Malta

60. Montenegro

61. Netherlands

62. Northern Ireland

63. North Europe

64. Norway

65. Poland

66. Portugal

67. Romania

68. Scandinavia

69. Scotland

70. Serbia

71. Silesia

72. Slovakia

73. Slovenia

74. South Europe

75. Spain

76. Sweden

77. Switzerland

78. Ukraine

79. United Kingdom

80. Upper Silesia

81. Vatican

82. Wales

83. West Europe

Cities/Urban Areas

84. Amsterdam

85. Antwerp

86. Athens

87. Augsburg

88. Avignon

89. Barcelona

90. Belfast

91. Belgrade

92. Berlin

93. Bern

94. Bratislava

95. Bruges

96. Brussels

97. Bucharest

98. Buchenwald

99. Budapest

100. Cambridge

101. Copenhagen

102. Dijon

103. Dresden

104. Dublin

105. Florence

106. Frankfurt

107. Genoa

108. (The) Hague

109. Helsinki

110. Innsbruck

111. Kohtla-Jarve (Latvia)

112. Krakow

113. Lisbon

114. Liverpool

115. Ljubljana

116. London

117. Luxembourg (City)

118. Madrid

119. Manchester

120. Marseille

121. Milan

122. Nicosia

123. Oslo

124. Paris

125. Prague

126. Reykjavik

127. Riga

128. Rome

129. Rotterdam

130. Sarajevo

131. Skopje

132. Sofia

133. Stockholm

134. Stuttgart

135. Tallinn

136. Tirane

137. Titograd

138. Valletta

139. Venice

140. Vienna

141. Vilnius

142. Warsaw

143. Zagreb

MAPPING EXERCISES

The following are three mapping exercises to improve your knowledge of the location of places, underscore why they are important, and clarify how they relate to each other. Some questions will ask you to locate places, compare maps, or fill in data; others will test your understanding of *why* you were asked to map the features that you did. Use the blank outline maps at the end of the chapter to complete these exercises. Additional blank outline maps can be found on the textbook's Web site: www.whfreeman.com/pulsipher4e.

1. River basins, population, and pollution

Rivers play an important role in the continued success of European economies. Significant rivers flow through several countries and provide valuable transportation access. Manufacturing districts grew up in areas that formerly provided important industrial materials. Unfortunately, these valuable rivers are also highly polluted. Using the map of Europe, complete the following:

- Label the following rivers: the Rhine and the Danube.
- Select a pattern to show the high-tech manufacturing and high quality/luxury goods manufacturing areas (Figure 4.20).
- Using the map of population density (Figure 4.11), shade the areas (in red) that have a population density of greater than 250 persons per square kilometer (greater than 650 persons per square mile).
- Locate and label all cities with 2 million people or more (Figure 4.11).
- Select a pattern to show the areas of high human impact on the land (Figure 4.33).

Questions

a. Using this mapped information, briefly describe the pattern and distribution of population.
b. Is population concentrated in any particular location (e.g., along the coasts, along major rivers, or in the interior)?
c. Is population concentrated near or away from manufacturing centers?
d. Summarize the relationship between the two rivers, population concentrations, manufacturing, and pollution.
e. Suggest some changes that will still allow the rivers to be used as major arteries for transport while at the same time permitting successful cleanup and good stewardship.

2. De-industrialization of Europe: High-tech manufacturing in the twenty-first century.

Figure 4.20 shows the geographic change of the industrial centers in Europe as high-tech production has become a major economic activity. Figure 4.3 (map of the European Union) lists the dates of each country's entry into the EU, and Figure 4.21 depicts the transportation network for this region. After a careful reading of the section on the EU and observation of the three maps, respond to the following questions.

a. How does the growth of the EU correlate with the shift to high-tech manufacturing?
b. What is the relationship between the transportation network and the growth of the high-tech centers?
c. To what extent is the EU policy of extending development into previously marginalized areas reflected in the maps showing centers of production?
d. What efforts by the EU are needed to extend the high tech production into even more of the marginalized areas? Give some examples of good practices that could benefit some of the newest EU members.

3. Women's empowerment in Europe

Some European countries have shown exemplary progress in bringing women into equally responsible positions in government and related socio-economic strata of society. However, these countries may not be representative of the entire European region.

- On the blank map of Europe, map GEM as listed in Table 4.1 (human well-being rankings) using graduated circles for each country.
- Map the welfare systems (Figure 4.32) using a different color for each of the different welfare systems.

Questions

a. Describe any patterns that suggest a correspondence between certain welfare regimes and women's overall empowerment in different European countries.

b. Based on your reading of the textbook, what do you think is the driving force in areas where women are highly empowered and actively engaged in political agendas?

c. After reading the textbook sections on gender and social welfare systems (along with reference to the section on subregions), suggest a country in Europe that could serve as a good role model for promoting women's empowerment. Give three reasons for why you selected this country.

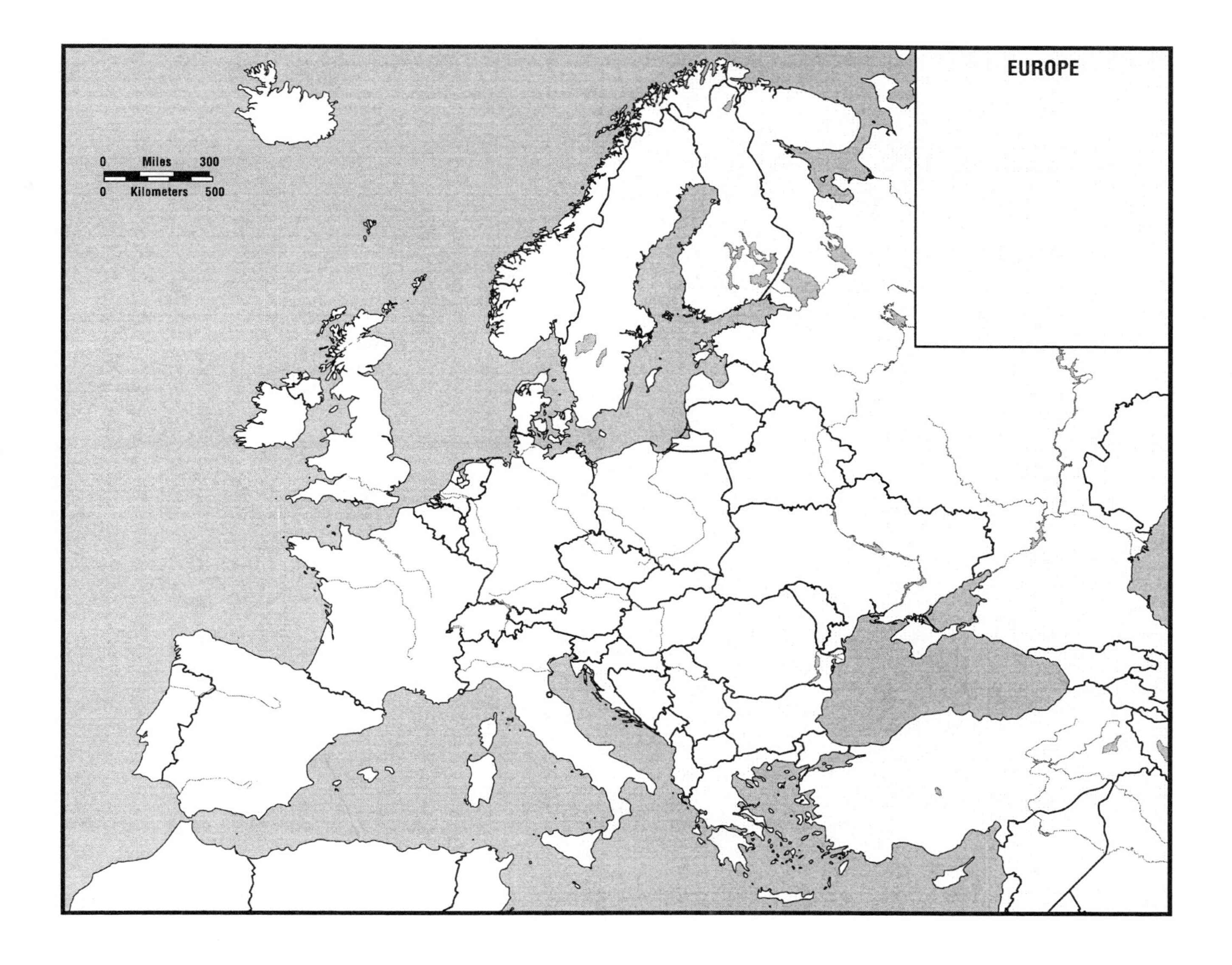
EUROPE
0 Miles 300
0 Kilometers 500

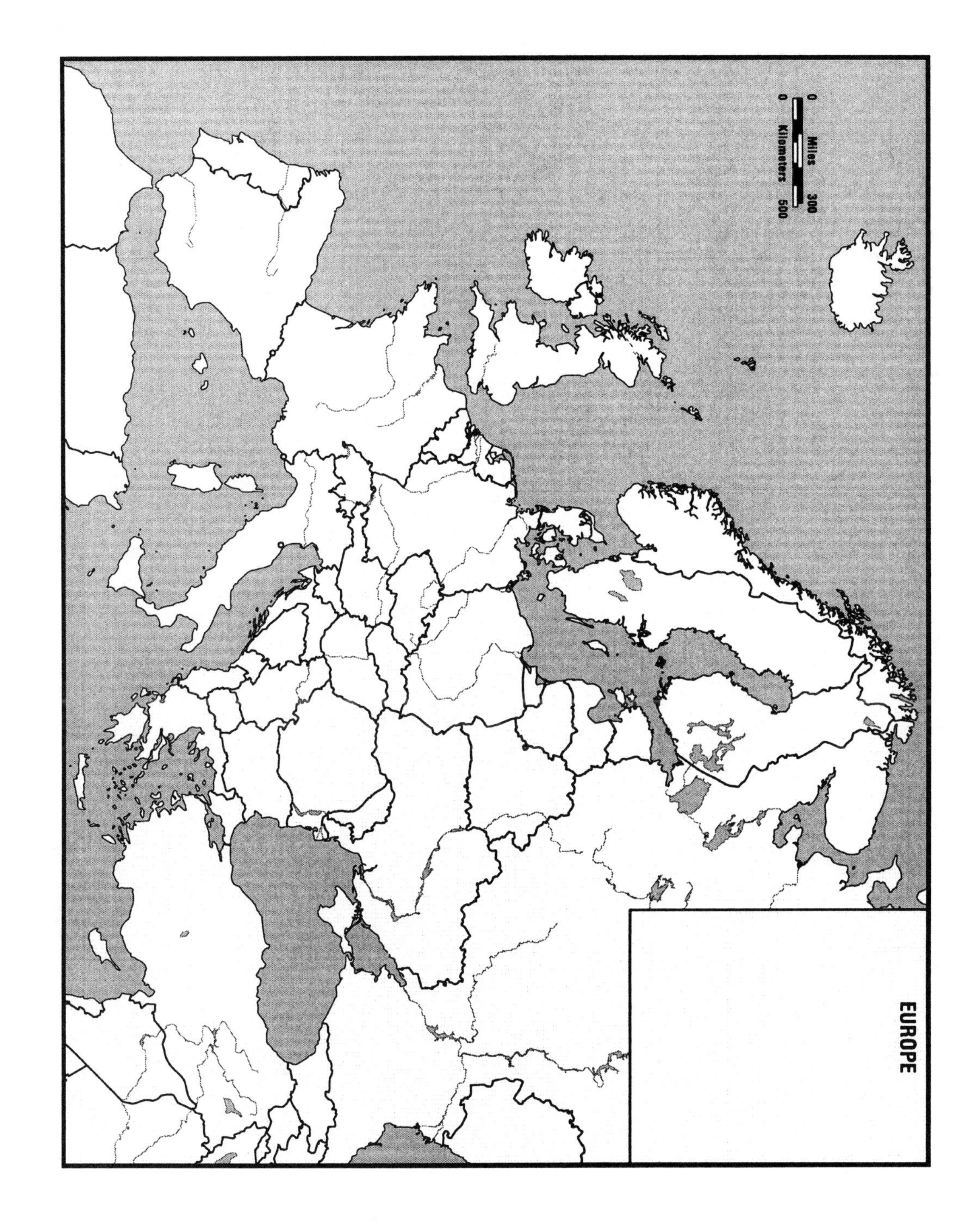
0 Miles 300
0 Kilometers 500
EUROPE

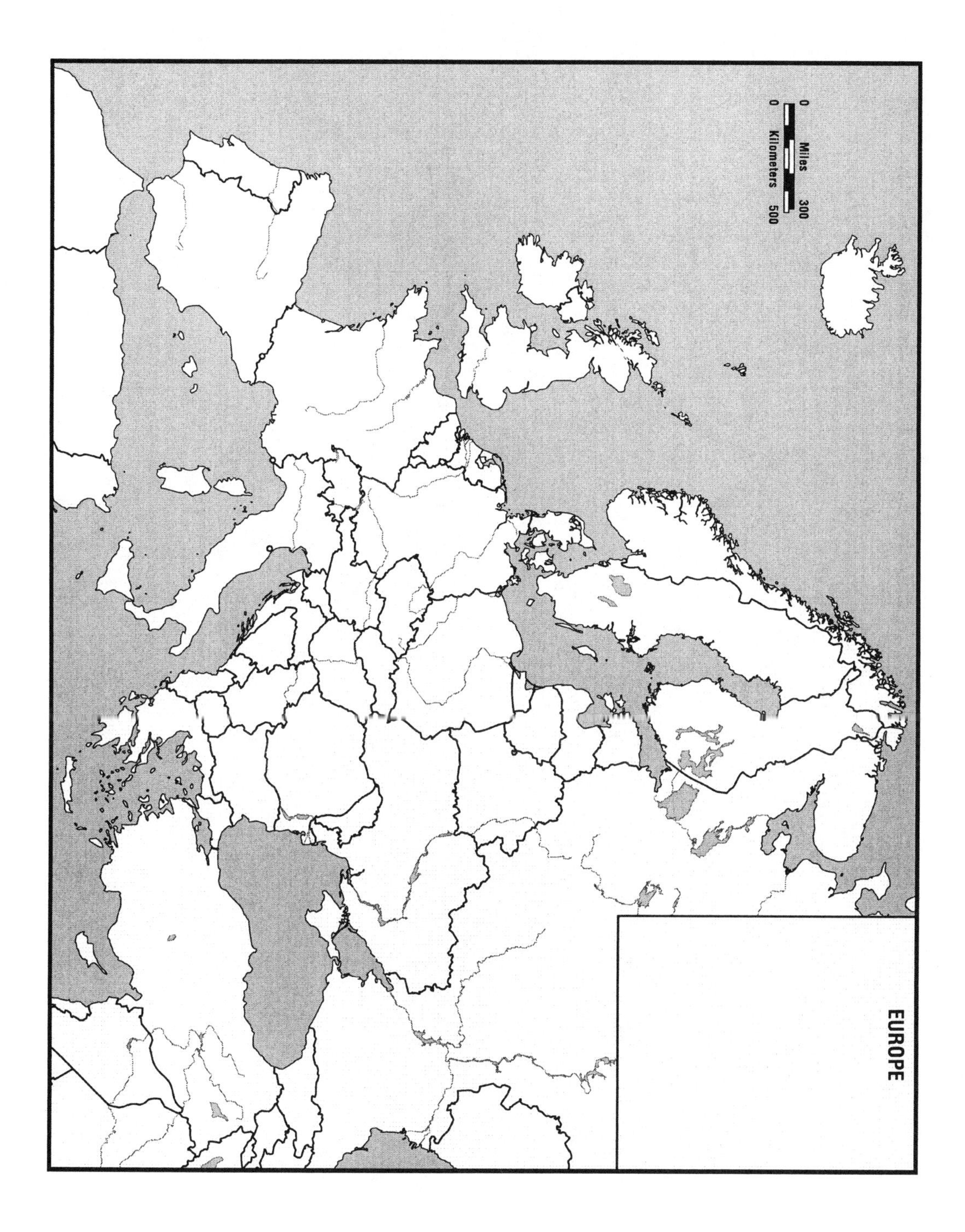
EUROPE
0 Miles 300
0 Kilometers 500

CHAPTER FIVE
Russia and the Newly Independent States

IMPORTANT PLACES

The following places are featured in the chapter. Make sure you can locate all of them on a map. A blank outline map of the region is provided at the end of this chapter; additional maps can be found on the textbook's Web site: www.whfreeman.com/pulsipher4e. Also, to prepare for exams, write a few important facts about each place in the space provided.

Physical Features

1. Altai Mountains

2. Amu Darya River

3. Amur River

4. Aral Sea

5. Arctic Ocean

6. Baltic Sea

7. Black Sea

8. Carpathian Mountains

9. Caspian Sea

10. Caucasus Mountains

11. Central Siberian Plateau

12. Gulf of Finland

13. Hindu Kush Mountains

14. Kamchatka Peninsula

15. Kolmya River

16. Lake Baikal

17. North European Plain

18. Ob River

19. Pacific Ocean

20. Pamir Mountains

21. Sakalin Island

22. Sea of Okhotsk

23. Siberia

24. Sikhote Alin

25. Syr Darya River

26. Tien Shan Mountains

27. Transcaucasia

28. Ural Mountains

29. Volga River

30. West Siberian Plain

31. White Sea

Regions/Countries/States/Provinces

32. Abkhazia

33. Armenia

34. Azerbaijan

35. Belarus

36. Caucasia (Caucasian Republics)

37. Central Asia

38. Chechnya

39. Dagestan

40. European Russia

41. Georgia

42. Ingushetiya

43. Kaliningrad

44. Kazakhstan

45. Kyrgyzstan

46. Moldova

47. Nagorno Karabakh

48. Russia

49. Tajikistan

50. Tatarstan

51. Turkmenistan

52. Tuva

53. Ukraine

54. Uzbekistan

Cities/Urban Areas

55. Angarsk

56. Ashkhabad

57. Astana

58. Baku

59. Bishkek

60. Bukhara

61. Chelyabinsk

62. Chernobyl

63. Chisinau

64. Donetsk

65. Dushanbe

66. Grozny

67. Irkutsk

68. Kazan

69. Kiev

70. Krasnoyarsk

71. Minsk

72. Moscow

73. Nakhodka

74. Norilsk

75. Novosibirsk

76. Odessa

77. Samarkand

78. Sevensk

79. St. Petersburg

80. Tashkent

81. Tbilisi

82. Vladivostok

83. Volgograd

84. Yekaterinburg

85. Yerevan

MAPPING EXERCISES

The following are three mapping exercises to improve your knowledge of the location of places, underscore why they are important, and clarify how they relate to each other. Some questions will ask you to locate places, compare maps, or fill in data; others will test your understanding of *why* you were asked to map the features that you did. Use the blank outline maps at the end of the chapter to complete these exercises. Additional blank outline maps can be found on the textbook's Web site: www.whfreeman.com/pulsipher4e.

1. Industry, pollution, and populations at risk

We can learn much about the populations at risk in this region by using maps to generate information.

- Construct a map that shows areas with population densities of 261 people or more per square mile (101 people or more per square kilometer) (Figure 5.13). A good symbolization would be shading using the darkest shade to represent highest population concentrations.
- Overlay areas of high industrial development (Figure 5.20). Categorize these by type of industrial region as designated in the legend for Figure 5.20.
- Overlay areas of oil and natural gas producing regions (Figure 5.21). Again use a unique symbol, such as cross-hatching.
- Finally, overlay areas of human impacts (Figure 5.30) using unique symbols for high impact and medium-high impact.

Questions

a. What appears to be a major source of employment for people in the densely populated areas?
b. What conclusions can you draw about environmental impacts on health? What are the governmental responsibilities for citizens' health?
c. What are three ways this region can cope with the environmental impacts of industrialization, resource extraction, and production?

2. Population issues in Russia and the newly independent states

By mapping demographic data by country, you should be able to arrive at some conclusions about population growth in this region.

- Using the CIA World Factbook at www.cia.gov/cia/publications/factbook, construct a map that portrays the birth rate for each country in this region. Use graduated circles to represent birth rates.

- Refer to Table 5.2 (human well-being rankings) to construct an overlay that depicts GDP per capita for each country. Be sure the lightest shade represents the lowest value and the darkest shade represents the highest value.
- Refer to Table 5.2 to construct another overlay that depicts the HDI for each country. Use diagonal lines (///) to indicate HDI. Use closely spaced lines to represent those countries with good HDI values (low numbers) and use widely spaced lines to represent those countries with poor (high number) HDI values.

Questions

a. Use this mapped information to help explain why some countries have a decreasing birth rate and why others have an increasing birth rate.
b. If GDP and HDI are not fully explanatory for some of the countries, suggest three other factors that might explain those countries' birth rates.

3. Water and oil: Critical resources that affect international relations in the Central Asian Republics

Control and access to headwaters or upstream waters of the major rivers of the Central Asian Republics may be a deciding factor in the continuation of cotton production in the region. Oil pipeline routes may change as newly independent countries develop international ties with countries that were never part of the former USSR.

- Label the Central Asian Republics and label their neighbors.
- Use the regional map (Figure 5.1) to label and outline the rivers (in blue) that supply water to Central Asia's irrigation projects. Shade in all the countries that are drained by these rivers.
- On Figure 5.7 (map of agriculture), observe the cash crops in the Central Asia valleys. These are most likely cotton crops. Shade these areas on your map.
- Use the oil and natural gas resource map (Figure 5.21) to cross-hatch the location of oil- and gas-producing areas in this region.
- Also from Figure 5.21, draw in red the current oil and gas pipelines that move these resources to shipping ports. Using a second color, suggest alternative routes that lie outside of the region for moving this oil to shipping ports.

Questions

a. Identify countries that will have to develop some basis for negotiating use of the water from these rivers. How will upstream countries have an advantage? What counterarguments or possible offers can the downstream countries (where the rivers drain last) provide in order to assure access to the water?
b. Make an argument for how the new pipelines you suggested would be the most plausible considering international relations in the region, international demand, security, and certainly environmental sensitivity.

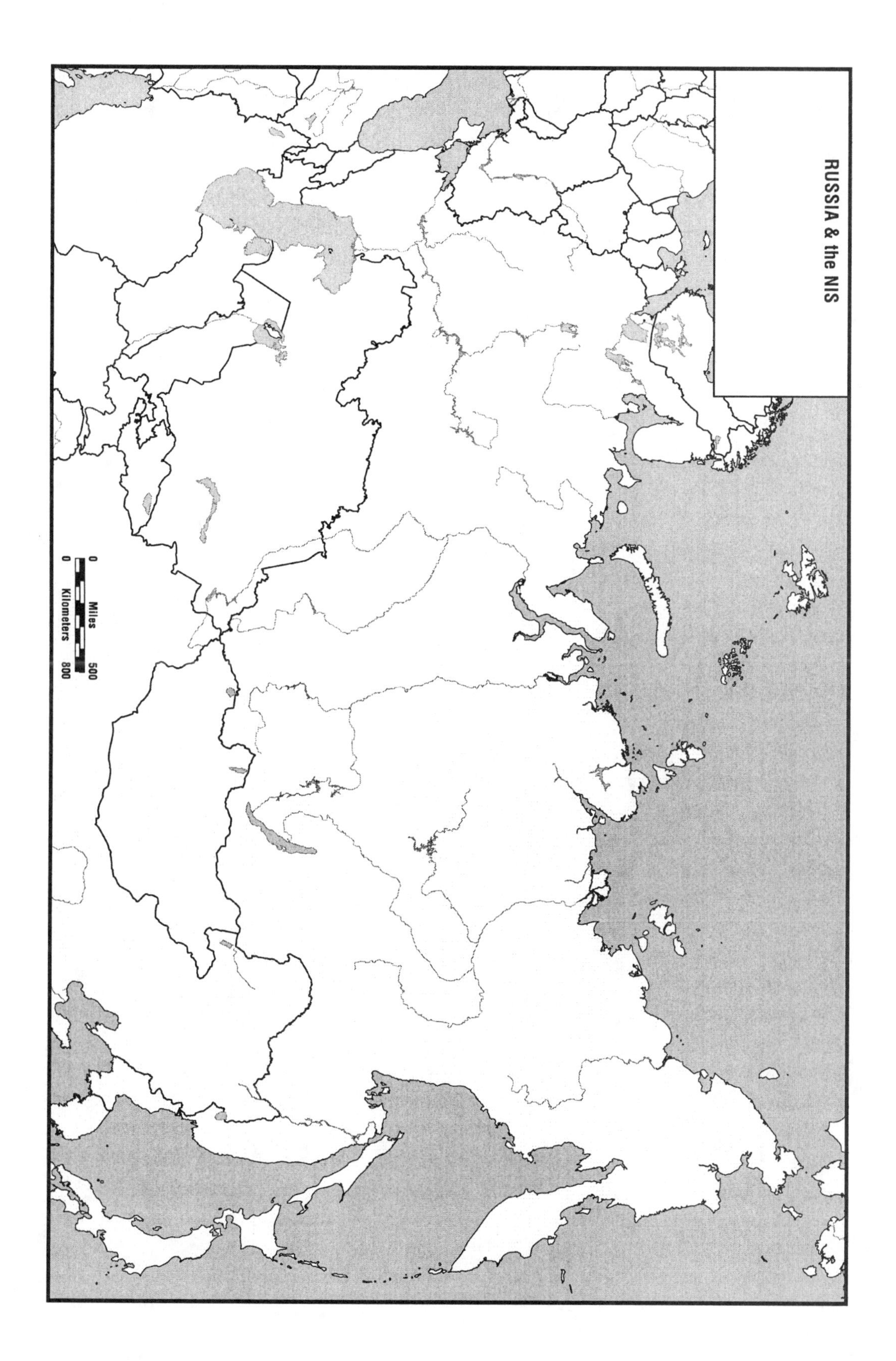
RUSSIA & the NIS
0 Miles 500
0 Kilometers 800

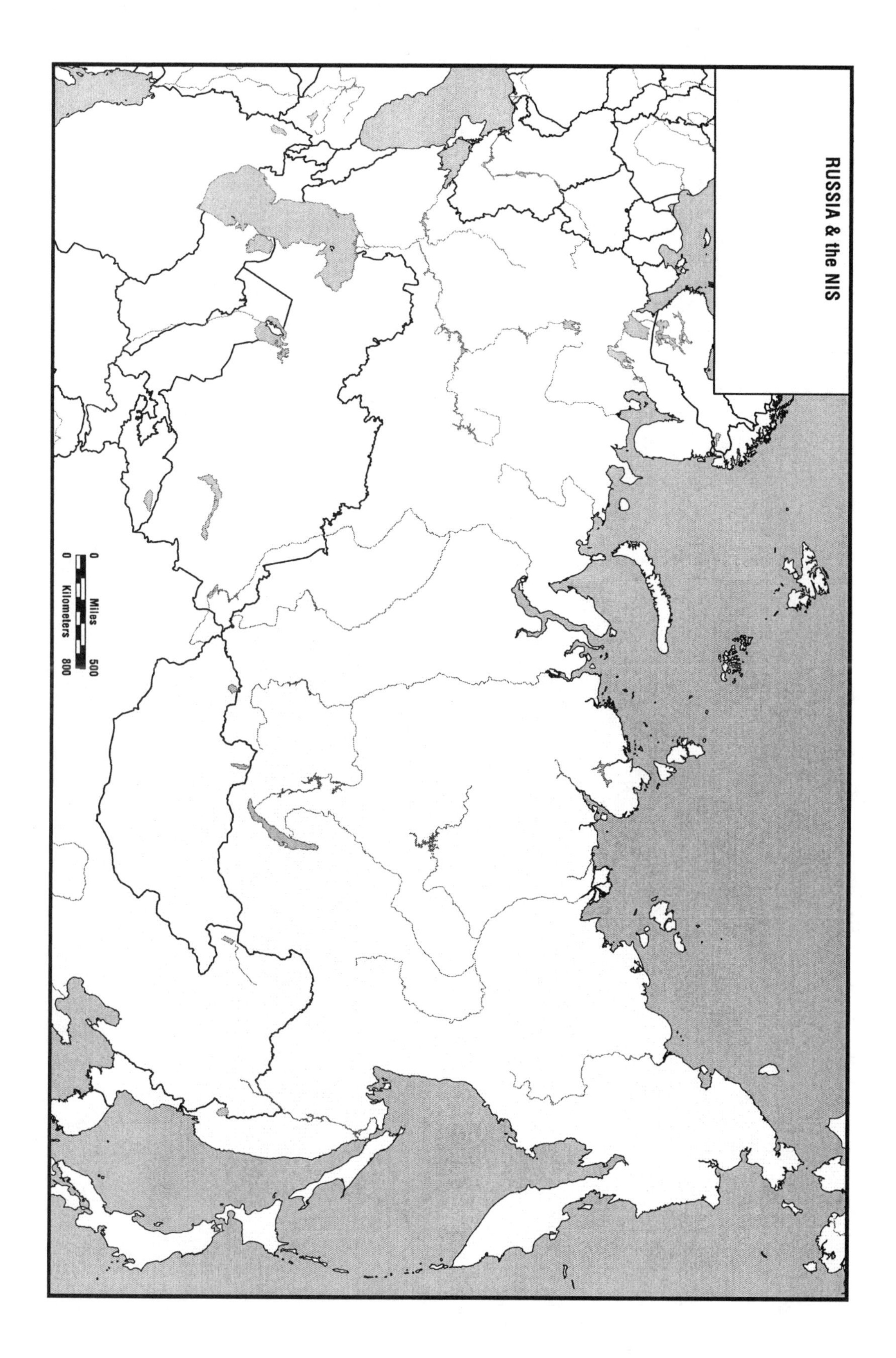
RUSSIA & the NIS
0 Miles 500
0 Kilometers 800

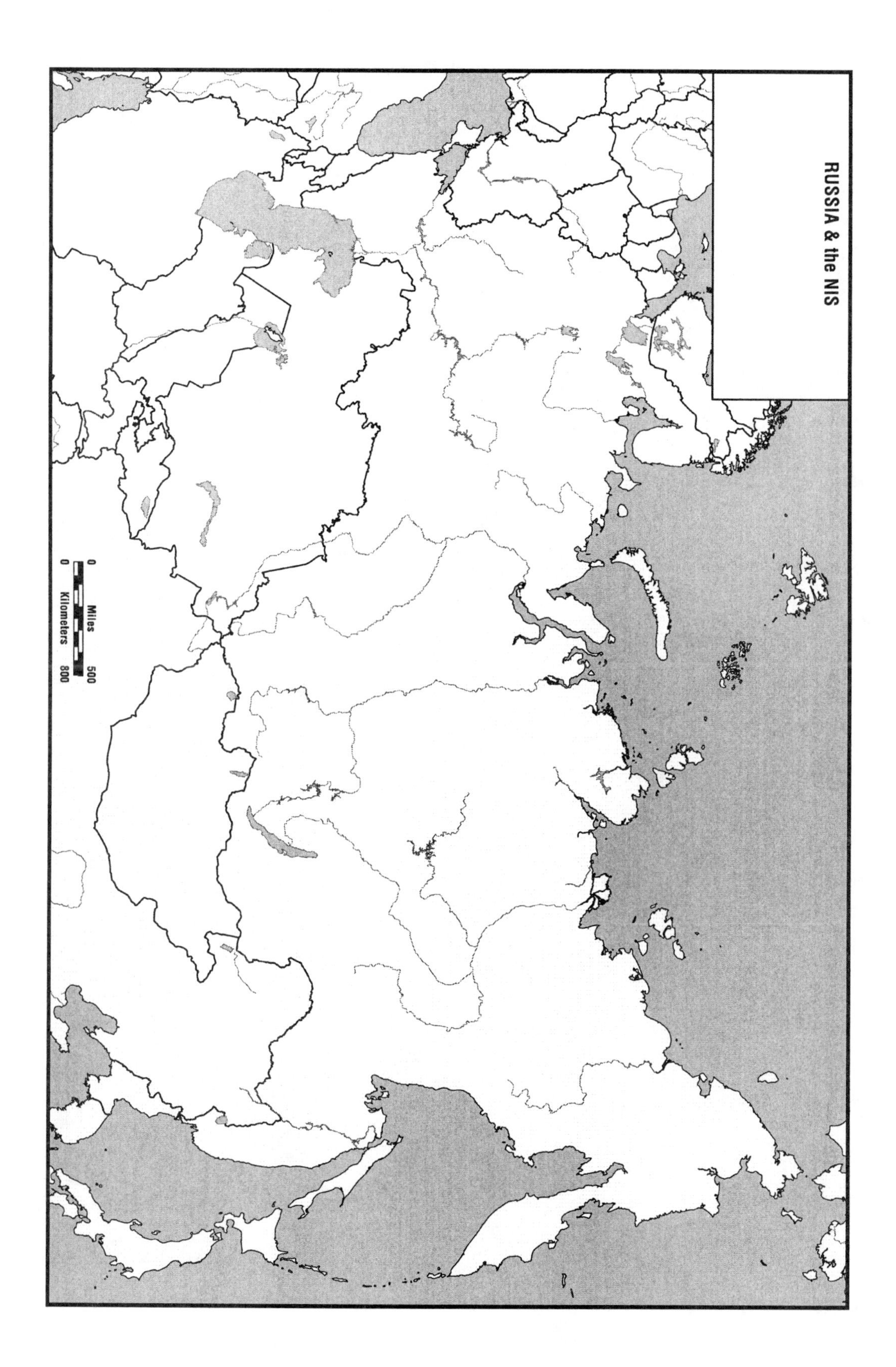
RUSSIA & the NIS
0 Miles 500
0 Kilometers 800

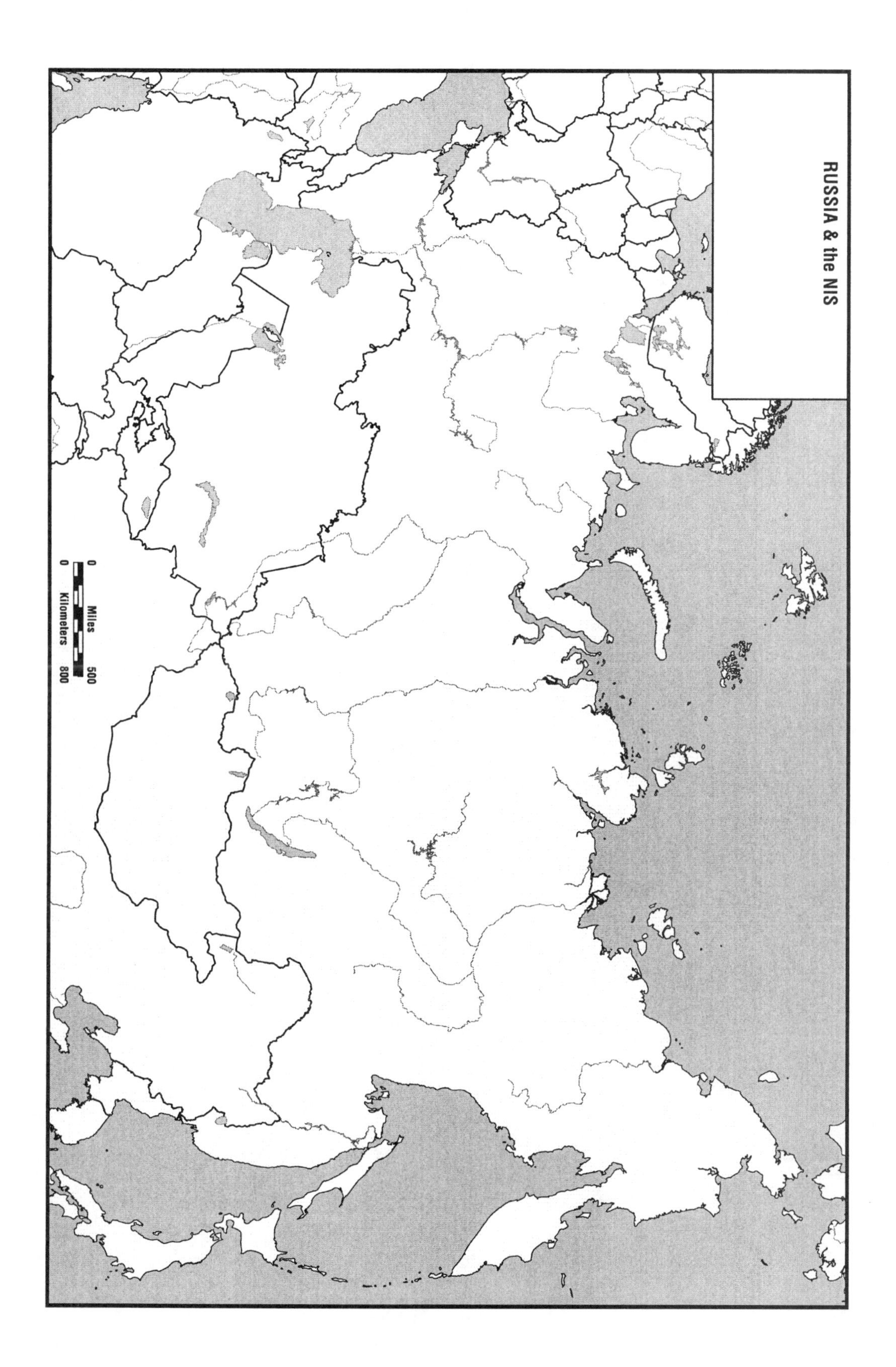
RUSSIA & the NIS
0 Miles 500
0 Kilometers 800

CHAPTER SIX
North Africa and Southwest Asia

IMPORTANT PLACES

The following places are featured in the chapter. Make sure you can locate all of them on a map. A blank outline map of the region is provided at the end of this chapter; additional maps can be found on the textbook's Web site: www.whfreeman.com/pulsipher4e. Also, to prepare for exams, write a few important facts about each place in the space provided.

Physical Features

1. Arabian Peninsula
2. Arabian Sea
3. Atlantic Ocean
4. Atlas Mountains
5. Black Sea
6. Caspian Sea
7. Dead Sea
8. Euphrates River
9. Gulf of Aden
10. Jordan River
11. Mediterranean Sea
12. Nile River
13. Persian Gulf
14. Red Sea
15. Rub'al Khali Desert
16. Sahara Desert

17. Tigris River

18. Zagros Mountains

Regions/Countries/States/Provinces

19. Algeria

20. Bahrain

21. Cyprus

22. Egypt

23. Gaza Strip

24. Golan Heights

25. Iran

26. Iraq

27. Israel

28. Jordan

29. Kuwait

30. Lebanon

31. Libya

32. The Maghreb

33. Morocco

34. Oman

35. Occupied (Palestinian) Territories

36. Qatar

37. Saudi Arabia

38. Sinai Peninsula

39. Sudan

40. Syria

41. Tunisia

42. Turkey

43. United Arab Emirates

44. West Bank

45. Western Sahara

46. Yemen

Cities/Urban Areas

47. Abu Dhabi

48. Algiers

49. Al Manamah

50. Amman

51. Ankara

52. Baghdad

53. Beirut

54. Cairo

55. Damascus

56. Doha

57. Gaza

58. Jerusalem

59. Khartoum

60. Kirkuk

61. Kuwait

62. Makkah (Mecca)

63. Medina

64. Muscat

65. Nicosia

66. Rabat

67. Riyadh

68. San'a

69. Tehran

70. Tripoli

71. Tunis

MAPPING EXERCISES

The following are three mapping exercises to improve your knowledge of the location of places, underscore why they are important, and clarify how they relate to each other. Some questions will ask you to locate places, compare maps, or fill in data; others will test your understanding of *why* you were asked to map the features that you did. Use the blank outline maps at the end of the chapter to complete these exercises. Additional blank outline maps can be found on the textbook's Web site: www.whfreeman.com/pulsipher4e.

1. Urbanization, female literacy, and opportunities for women

Rural women are often less secluded because they have many tasks to perform out of the home. However, women tend to have more education and job opportunities in urban areas.

- From the map of urban population (Figure 6.18), shade countries from light to dark that have the following percent of urbanization: 20-39 percent; 40-59 percent; 60-79 percent; and 80 percent or more.
- Using the table of human well-being rankings (Table 6.2), use a graduated symbol to represent the female literacy rate for all the countries in the region. For example, use a small triangle for 0-50 percent, a medium triangle for 51-75 percent, and a large triangle for 76-100 percent
- You will also need to refer to the map of women who are wage-earning workers

(Figure 6.15) and the map of restrictions placed on women (Figure 6.24).

Questions

a. What relationship would you expect between a country's urbanization and: 1) the number of women who are earning wages, and 2) the restrictions placed on women? Why would you expect these relationships? Where do you see them?
b. What relationship would you expect to see between female literacy rates and: 1) the percentage of women who are earning wages, and 2) the restrictions placed on women? Where do you see these relationships?
c. Choose two countries that stand out as contradictory to what you would expect. Explain why this might be the case.

2. Oil wealth and human well-being

Oil wealth in the region is not evenly distributed. Many live in poverty with a less than desirable quality of life, while a few are extremely wealthy.

- Using Figure 6.27 (map of economic issues), carefully shade (in pencil) the oil and gas-producing areas.
- Also using Figure 6.27, cross-hatch (///) the countries that are OPEC members.
- Using Table 6.2 (human well-being rankings), write (inside the country border) the GDP per capita and HDI for each country in the region.

Questions

a. List the countries that have oil and gas production. What is the general relationship between GDP per capita and oil and gas production? Are there any anomalies? If so, explain why.
b. For each oil and gas-producing country, compare its GDP per capita to its HDI. Overall, in countries where GDP per capita is high, does HDI also rank high (remember that a high HDI ranking appears in the table as a low number)? Is it what you expected? Why or why not?
c. Overall, how is oil *positively* affecting the region's people, the economy, and politics? How is it *negatively* affecting the region's people, the economy, and politics?

3. The importance of (clean) water

Most people in this region understand the importance, even the necessity, of living near sources of water; thus, it is vital that these water sources are clean. Refer to the population density map (Figure 6.14) and examine patterns of population density.

Questions

a. Examining the map, how would you describe the pattern of population distribution? Is it concentrated in specific locations (e.g., coastal or interior; near water or deserts)?
b. Given the current distribution, where do you think the rapidly increasing numbers of people will live?
c. Next, examine the maps of human impact (Figure 6.35). Based on where you predicted the growing population would live, how might this growth be problematic? How might it further increase water shortages and desertification?

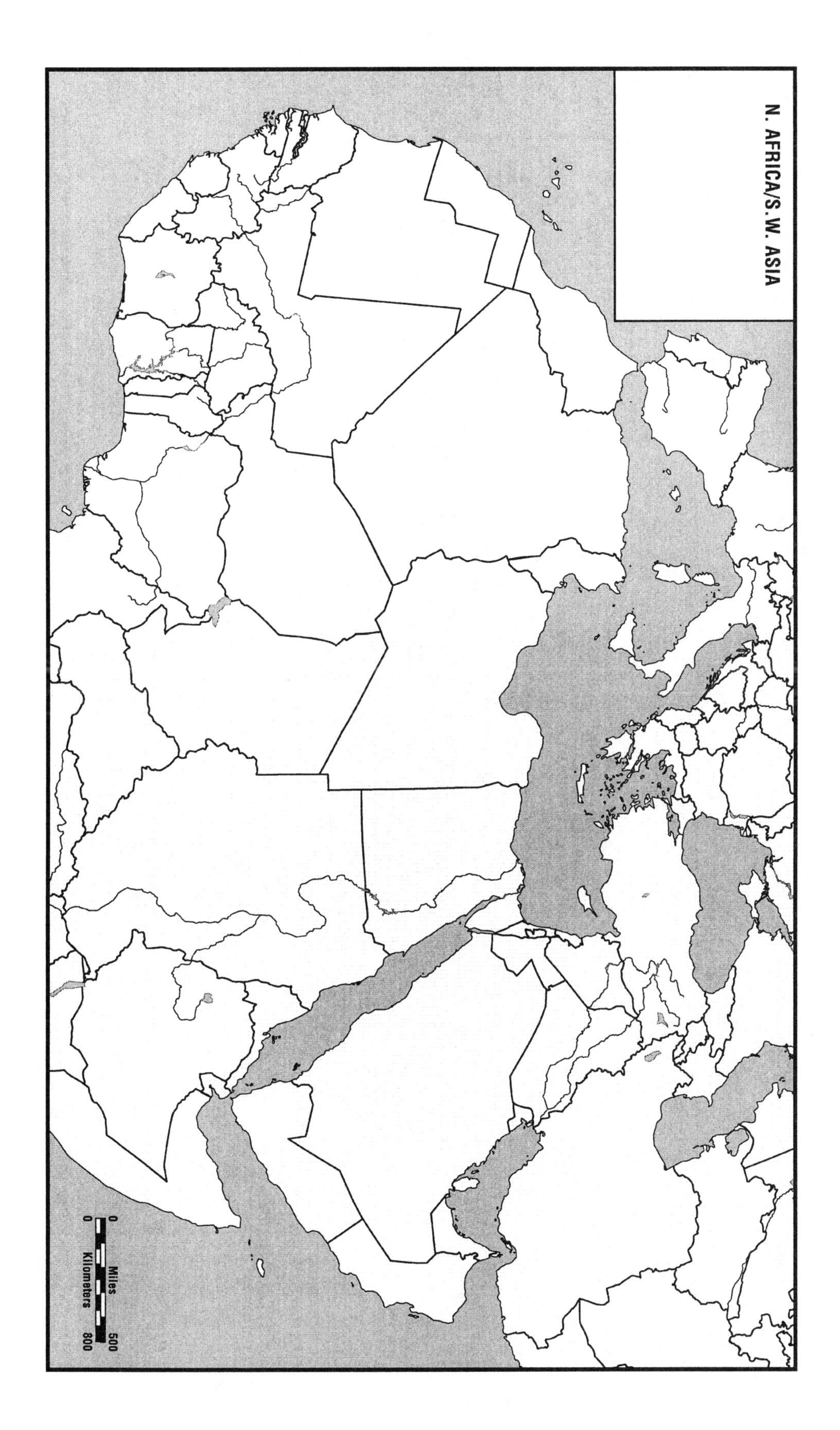
N. AFRICA/S.W. ASIA
0
Miles
500
0
Kilometers
800

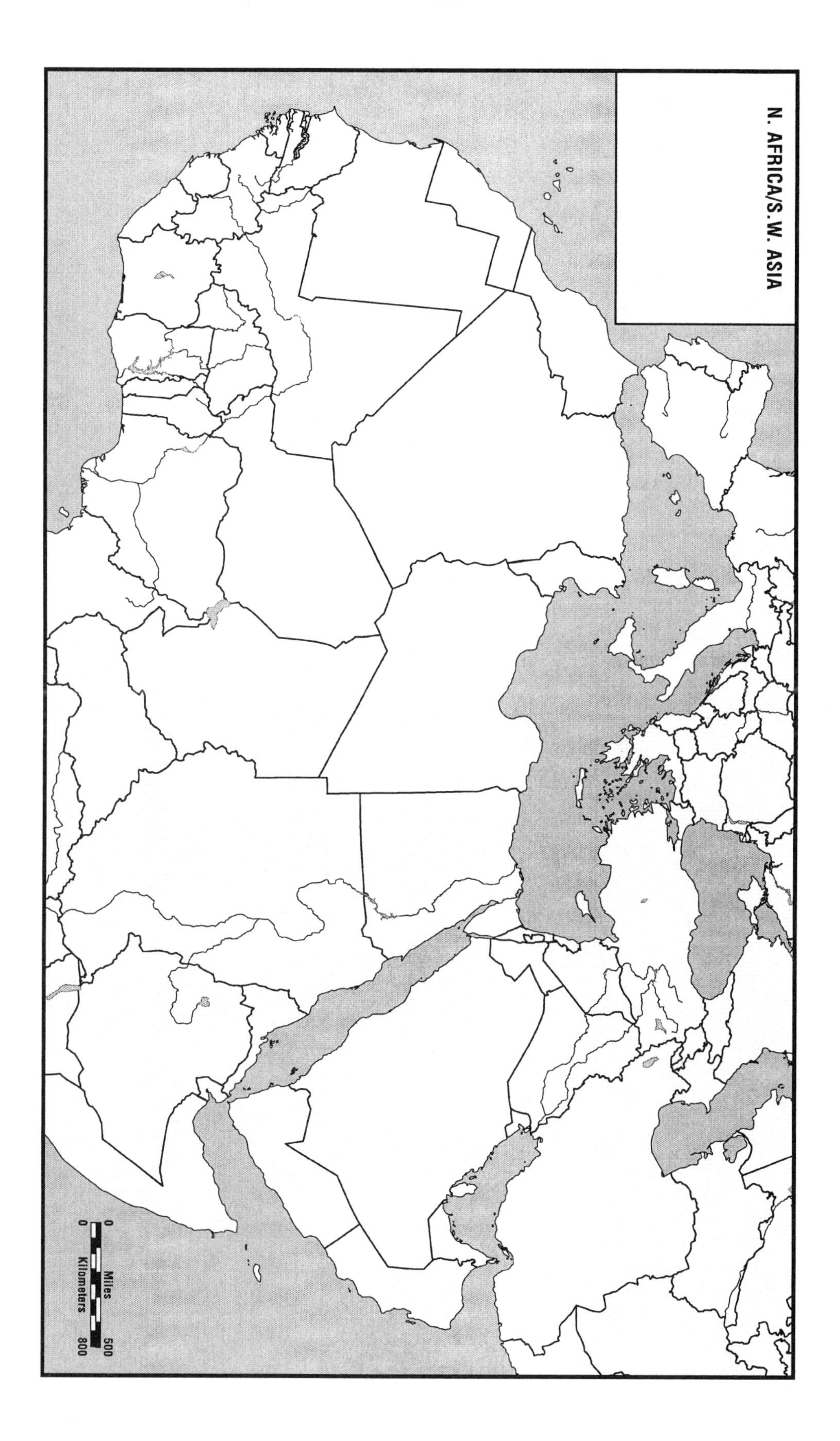
N. AFRICA/S.W. ASIA
0 Miles 500
0 Kilometers 800

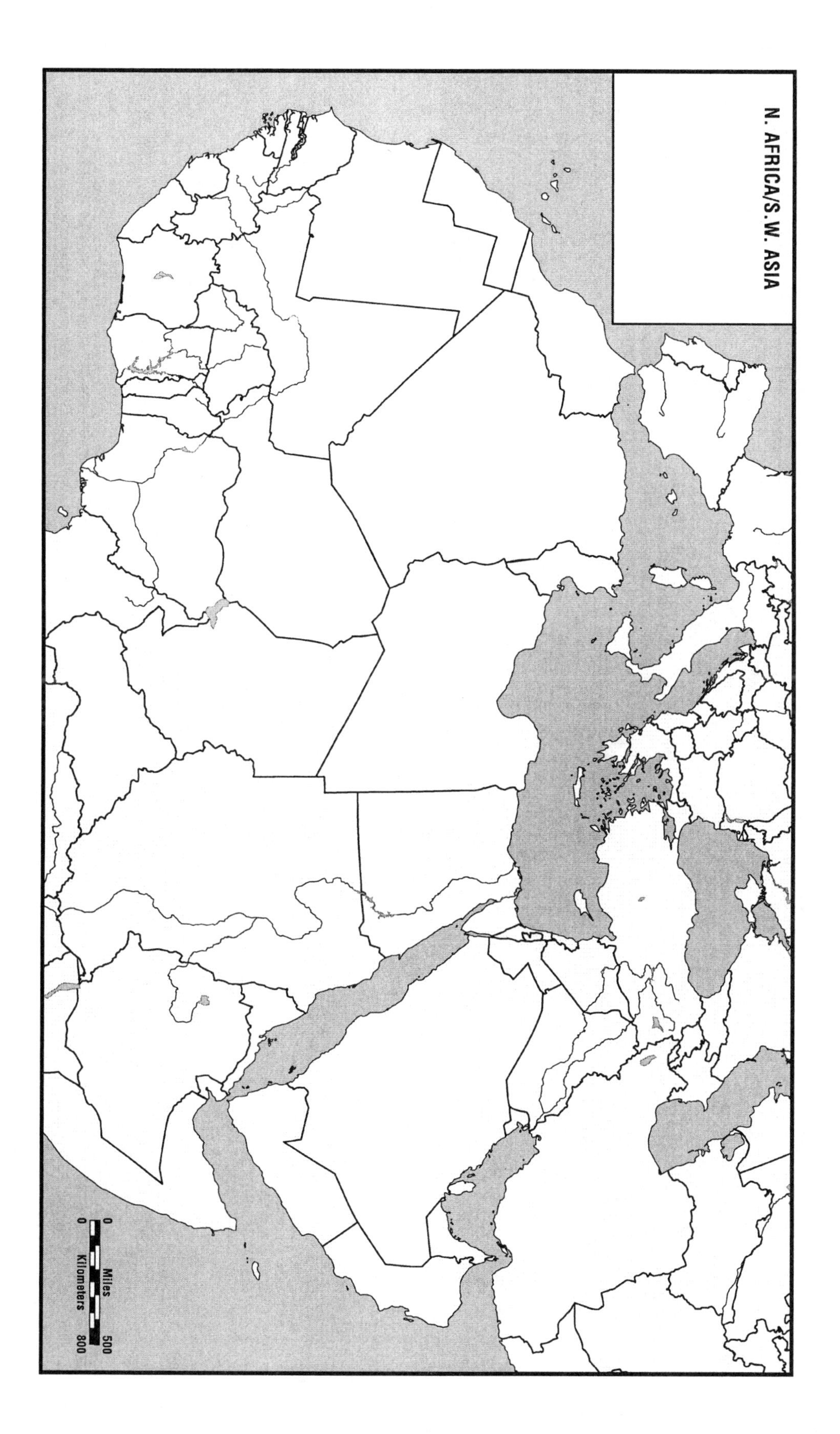
N. AFRICA/S.W. ASIA
0
Miles
500
0
Kilometers
800

CHAPTER SEVEN
Sub-Saharan Africa

IMPORTANT PLACES

The following places are featured in the chapter. Make sure you can locate all of them on a map. A blank outline map of the region is provided at the end of this chapter; additional maps can be found on the textbook's Web site: www.whfreeman.com/pulsipher4e. Also, to prepare for exams, write a few important facts about each place in the space provided.

Physical Features

1. Atlantic Ocean
2. Cape of Good Hope
3. Congo Basin
4. Congo (Zaire) River
5. Ethiopian Highlands
6. Great Rift Valley
7. Horn of Africa
8. Indian Ocean
9. Kalahari Desert
10. Lake Chad
11. Lake Malawi
12. Lake Tanganyika
13. Lake Victoria
14. Mount Kenya
15. Mount Kilimanjaro
16. Namib Desert

17. Niger River

18. Orange River

19. Red Sea

20. Sahara Desert

21. Sahel

22. Zambezi River

Regions/Countries/States/Provinces

23. Angola

24. Benin

25. Botswana

26. Burkina Faso

27. Burundi

28. Cameroon

29. Cape Verde Islands

30. Central African Republic

31. Chad

32. Comoros

33. Congo (Brazzaville)

34. Congo (Kinshasa)

35. Côte d'Ivoire

36. Djibouti

37. Equatorial Guinea

38. Eritrea

39. Ethiopia

40. Gabon

41. Gambia

42. Ghana

43. Guinea

44. Guinea-Bissau

45. Kenya

46. Lesotho

47. Liberia

48. Madagascar

49. Malawi

50. Mali

51. Mauritania

52. Mauritius

53. Mozambique

54. Namibia

55. Niger

56. Nigeria

57. Réunion

58. Rwanda

59. São Tomé and Principe

60. Senegal

61. Seychelles

62. Sierra Leone

63. Somalia

64. South Africa

65. Swaziland

66. Tanzania

67. Togo

68. Uganda

69. Zambia

70. Zimbabwe

Cities/Urban Areas

71. Abidjan

72. Abuja

73. Accra

74. Addis Ababa

75. Antananarivo

76. Asmera

77. Bamako

78. Bangui

79. Banjul

80. Bissau

81. Brazzaville

82. Bujumbura

83. Cape Town

84. Conakry

85. Dakar

86. Dar es Salaam

87. Djibouti

88. Freetown

89. Gaborone

90. Harare

91. Johannesburg

92. Kampala

93. Kigali

94. Kinshasa

95. Lagos

96. Libreville

97. Lilongwe

98. Lomé

99. Luanda

100. Lusaka

101. Malabo

102. Maputo

103. Maseru

104. Mbabane

105. Mogadishu

106. Mombasa

107. Monrovia

108. Moroni

109. Nairobi

110. N'Djamena

111. Niamey

112. Nouakchott

113. Ouagadougou

114. Port Louis

115. Porto-Novo

116. Praia

117. Pretoria

118. São Tomé

119. Victoria

120. Windhoek

121. Yamoussoukro

122. Yaounde

MAPPING EXERCISES

The following are three mapping exercises to improve your knowledge of the location of places, underscore why they are important, and clarify how they relate to each other. Some questions will ask you to locate places, compare maps, or fill in data; others will test your understanding of *why* you were asked to map the features that you did. Use the blank outline maps at the end of the chapter to complete these exercises. Additional blank outline maps can be found on the textbook's Web site: www.whfreeman.com/pulsipher4e.

1. Population growth, education, and HIV-AIDS

HIV-AIDS is the most severe public health problem in sub-Saharan Africa. It is having dramatic effects on population growth and life expectancy patterns. However, education is beginning to play a changing role in the spread and control of HIV-AIDS.

- Shade those countries that have 5 percent or more of adults with HIV-AIDS (Figure 7.17). Use a lighter shade for countries with 5-15 percent and a darker shade for countries with 15-34 percent. Label these countries.
- From the current year's "World Population Data Sheet," found at www.prb.org, look up the countries you labeled. Draw a hatch pattern over the countries that have over 2 percent rate of natural increase.
- Finally, use a graduated symbol (e.g., small to large circles), and draw a symbol in each of these countries to illustrate its female literacy (Table 7.1).

Questions

a. What is the current relationship between HIV-AIDS and population rate of natural increase? Explain why this is the case.
b. If the situation stays the same, how will HIV-AIDS affect life expectancy and population growth of this region 20 years from now? How could this situation be changed?
c. Which of these countries do you think will experience the greatest *decrease* in population growth rate? Why?
d. Examining the literacy data, how might increased education affect countries that have high HIV-AIDS rates? How will it affect those with high growth rates? Explain why. Some countries with high HIV-AIDS rates *do* have high rates of literacy; why do you think they still have high rates of HIV-AIDS?

2. Democracy and the provision of basic needs

In examining the daily suffering of many Africans, many wonder if African governments should meet their citizens' basic needs for food, shelter, and health care *before* they open up to a democratic form of government.

- Using Figure 7.35, shade the countries that had democratically elected governments in 1970. Choose a different color and shade the additional countries that had democratically elected governments in 2006. Cross hatch all countries that had democratically elected governments in 1970, but *not* in 2006. Label all the countries you shaded.
- Use graduated symbols to illustrate the Human Development Index as either "low," "medium," or "high" for all countries (Table 7.1).

Questions

a. Based on your map, which meets basic needs better, democratic or undemocratic governments? Why do you think this is the case? What can you interpret about the connection between HDI and democracy for those who had democratic elections in 1970 but *not* in 2006?
b. In light of the suffering of Africa's people, explain whether you think African governments should meet basic needs like food, shelter, and health care before they focus on achieving democracy.

c. What are some of the difficulties governments face in providing for citizens?
   - How can countries provide for needs when they have a very limited tax base from which to draw?
   - Many of these countries were established as democracies at independence, yet authoritarian presidents have caused major problems. How do you make a corrupt dictator take care of the basic needs of the people?

d. Explain if you think Africa would be better off left alone, or should other countries help provide basic needs for Africans?

3. The limits on carrying capacity

Carrying capacity depends on physical factors, including water supply and quality, soil condition, and disease, as well as cultural, social, economic, and political factors, including agriculture, wealth, and political unrest.

- Shade (in yellow) the countries with 5 percent or more of their population infected with HIV (Figure 7.17).
- Choose a symbol to draw in each country with GDP per capita less than $1000 (Table 7.1).
- Using the regional map of sub-Saharan Africa (Figure 7.1), outline (in light brown) and label the Kalahari, Namib, and Sahara deserts.
- Using Figure 7.1, trace and label the major rivers in the region with a heavy blue line: Niger, Orange, Congo (Zaire), Nile, White Nile, Blue Nile, and Zambezi.
- Using the map of human impact (Figure 7.45), shade areas that have high impact and medium-high impact on the land.

Questions

a. Make a list of at least five countries on your map that appear to have the potential for high carrying capacity (i.e., low human impact on the land, not in a desert, low percent of population with HIV-AIDS, and a relatively high GDP per capita).
   - From the current year's "World Population Data Sheet," found at www.prb.org, add their rate of natural increase to your list.
   - Is rapid population growth occurring? What implications might this population change have on these countries?

b. Next, from the "World Population Data Sheet," make a list of the ten fastest growing countries in sub-Saharan Africa. Label these ten countries on your map.
   - Examining your map, does the carrying capacity of each of these ten rapidly growing countries appear to be high or low? Write high or low next to each.
   - What implications might this rapid growth have on the countries that already have a low carrying capacity?

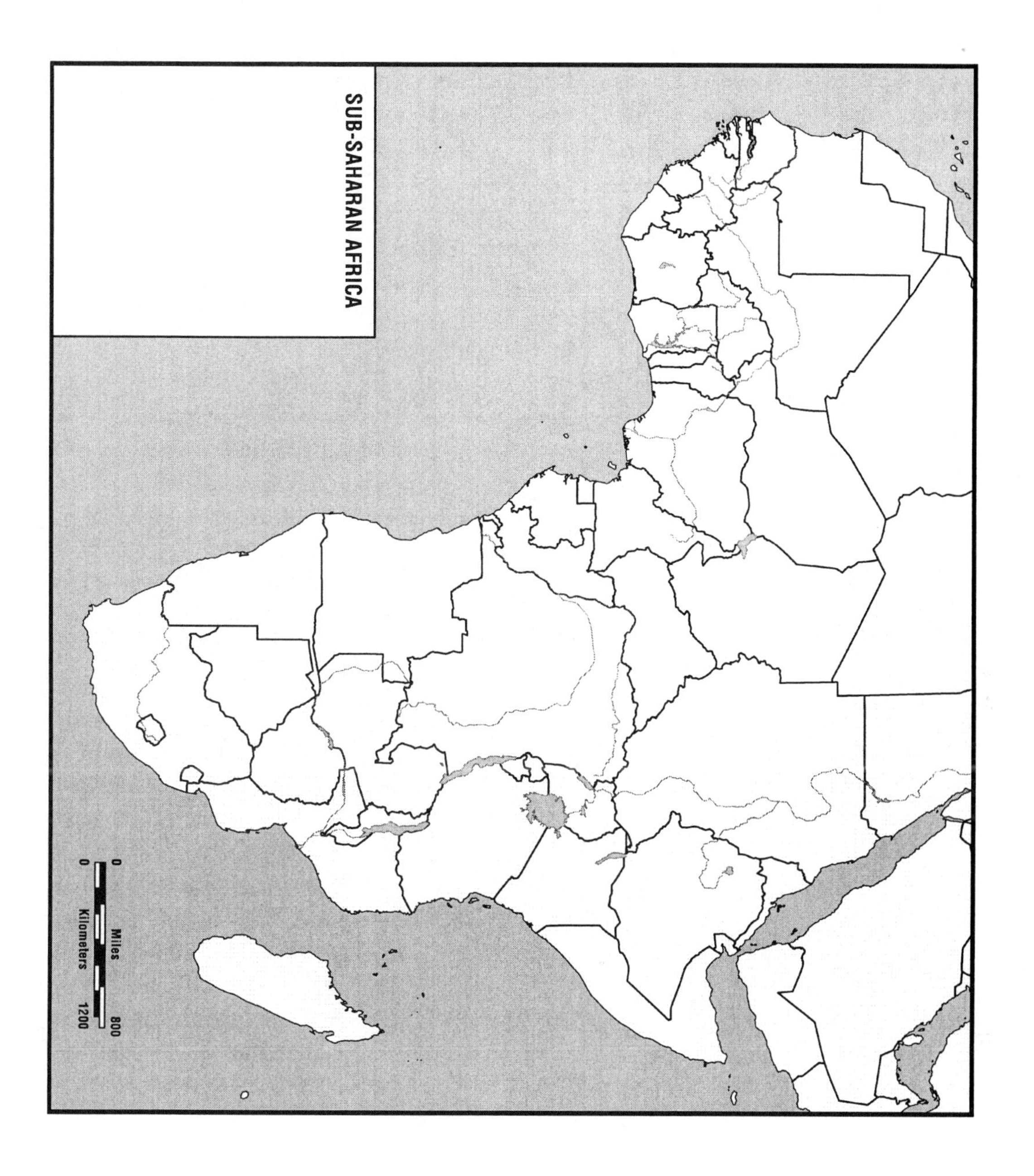
SUB-SAHARAN AFRICA
0 Miles 800
0 Kilometers 1200

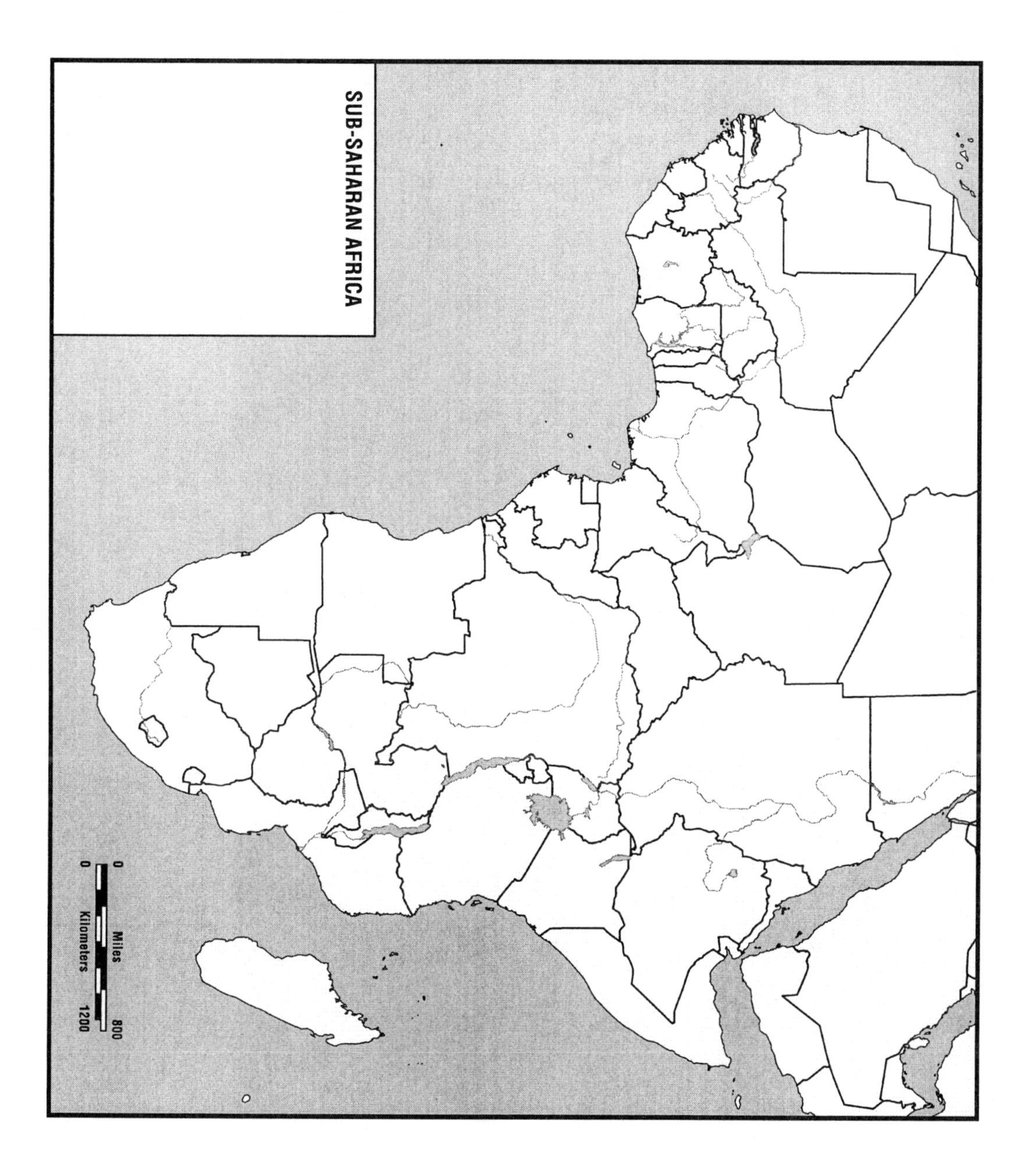
SUB-SAHARAN AFRICA
0 Miles 800
0 Kilometers 1200

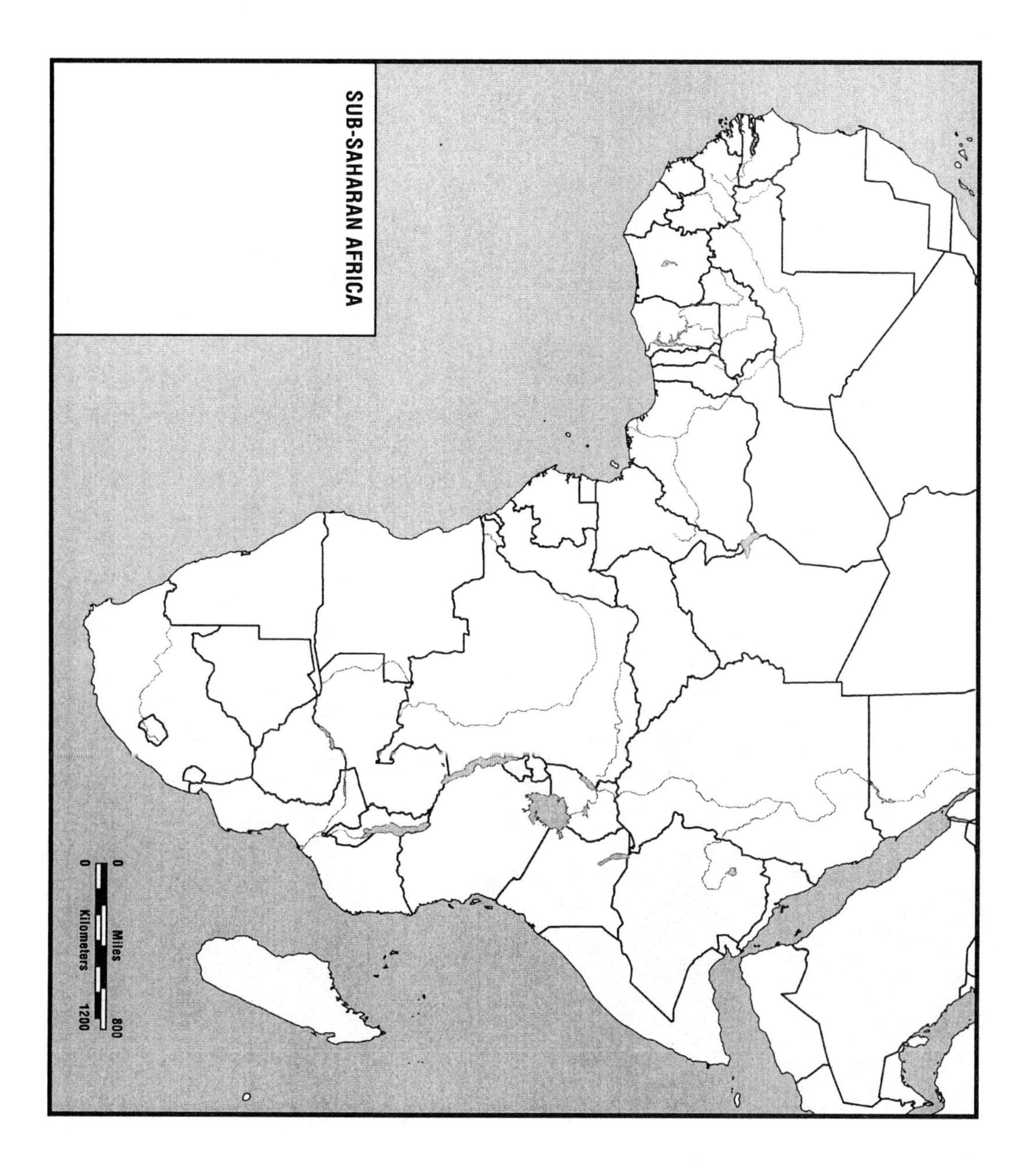
SUB-SAHARAN AFRICA
0
Miles
800
0
Kilometers
1200

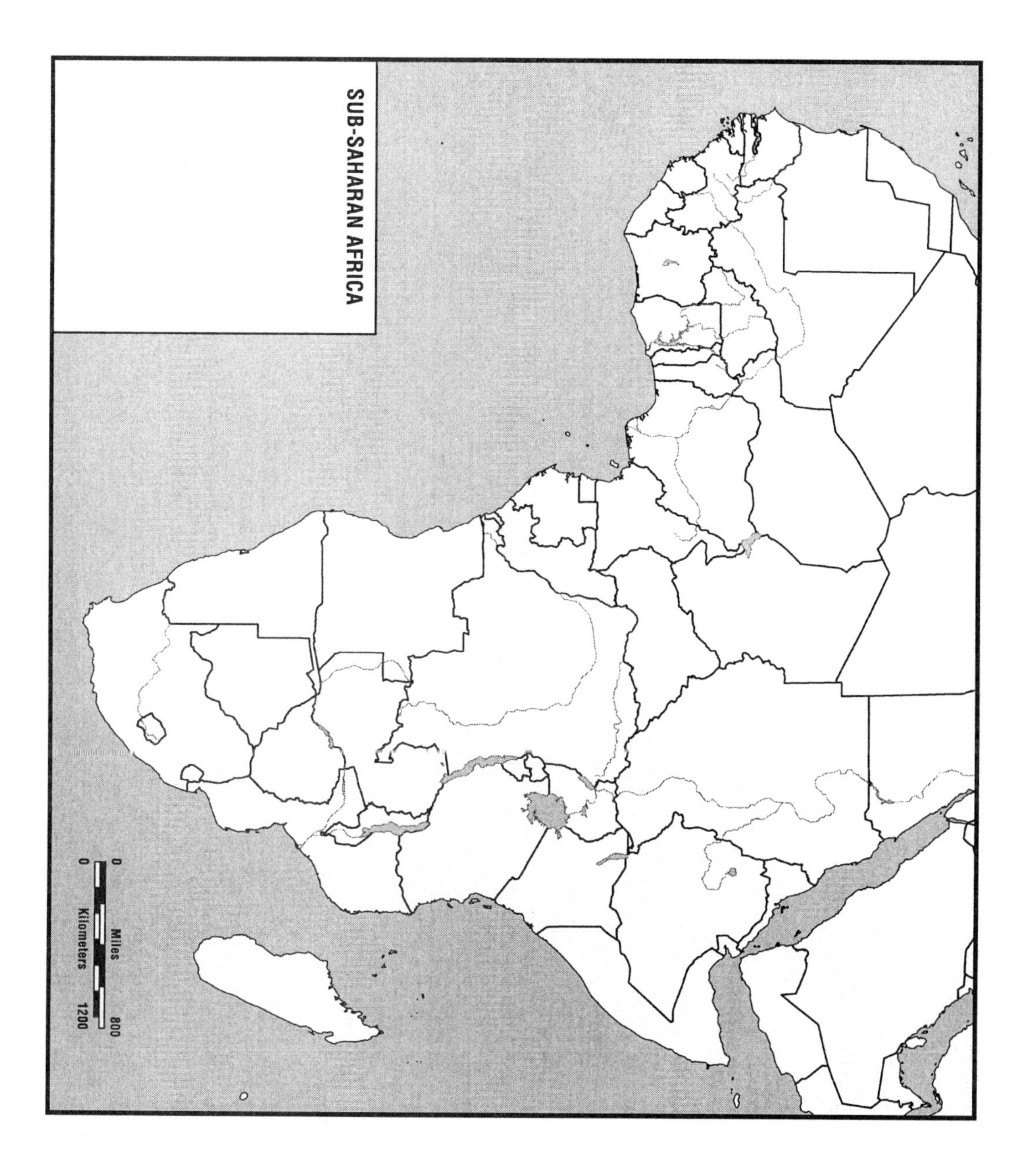
SUB-SAHARAN AFRICA
0
Miles
800
0
Kilometers
1200

CHAPTER EIGHT
South Asia

IMPORTANT PLACES

The following places are featured in the chapter. Make sure you can locate all of them on a map. A blank outline map of the region is provided at the end of this chapter; additional maps can be found on the textbook's Web site: www.whfreeman.com/pulsipher4e. Also, to prepare for exams, write a few important facts about each place in the space provided.

Physical Features

1. Arabian Sea

2. Bay of Bengal

3. Brahmaputra River

4. Deccan Plateau

5. Eastern Ghats

6. Ganga Plain

7. Ganga River

8. Himalayas

9. Hindu Kush

10. Indian Ocean

11. Indus River

12. Narmada River

13. Nilgiri Hills

14. Western Ghats

Regions/Countries/States/Provinces

15. Afghanistan

16. Ahraura

17. Bangladesh

18. Bhutan

19. Gujarat

20. India

21. Joypur

22. Kashmir

23. Kerala

24. Maldives

25. Nepal

26. Pakistan

27. Punjab

28. Sri Lanka

29. Uttar Pradesh

30. West Bengal

Cities/Urban Areas

31. Chennai (Madras)

32. Colombo

33. Delhi/New Delhi

34. Dhaka

35. Islamabad

36. Kabul

37. Kathmandu

38. Kolkata (Calcutta)

39. Mumbai (Bombay)

40. Thimphu

41. Varanasi (Benares)

MAPPING EXERCISES

The following are three mapping exercises to improve your knowledge of the location of places, underscore why they are important, and clarify how they relate to each other. Some questions will ask you to locate places, compare maps, or fill in data; others will test your understanding of *why* you were asked to map the features that you did. Use the blank outline maps at the end of the chapter to complete these exercises. Additional blank outline maps can be found on the textbook's Web site: www.whfreeman.com/pulsipher4e.

1. Relationship between landforms and population

South Asia has some of the most spectacular landforms and important rivers on earth. Some of these physical features make the land more habitable than others.

- From the map of the region (Figure 8.1), draw the outline, and label the Deccan Plateau.
- Draw (with ^^^) and label the location of these mountain ranges: Eastern Ghats, Western Ghats, Himalayas, and Hindu Kush (Figure 8.1).
- Draw a heavy blue line over these rivers: Brahmaputra, Ganga, Indus, and Narmada (Figure 8.1). Label each of these rivers.
- Shade the areas that have over 1300 persons per square mile (over 500 people per square kilometer) (Figure 8.13).

Questions

a. Based on the information you mapped, is population concentrated in specific locations (i.e., coastal or interior, along rivers or mountain ranges, etc.)? If so, where?
b. What is the general relationship between population and each of the different landforms: plateaus, mountains, and rivers? Explain the reasons for each.

2. How does climate affect people's lives and location?

This region has a great variety of regional climates, ranging from tropical monsoon to deserts. These physical conditions can have major impacts on living conditions.

- Shade in the general location of temperate climates (in green) and the arid and semiarid climates (in orange) (Figure 8.5).

- Use a hatch pattern (///) to show where population density is over 1300 persons per square mile (over 500 people per square kilometer) (Figure 8.13).
- Use the opposite hatch pattern (\\\) to show the areas with the heaviest monsoon rains (Figure 8.4).

Questions

a. Based on the information you mapped, what is the general relationship between population and each of the two climate types? Why is this the case for each?

b. Take a look at the areas where the hatch patterns cross (high population density *and* heavy monsoon rains). List at least three positive and three negative effects of monsoon rains on the people in these densely populated areas.

3. Female literacy and population growth rates

The overall status of women in South Asia is low; however, their relative well-being varies geographically. Freeing women from purdah encourages lower fertility rates and allows women to improve their own educational attainment and overall well-being, as well as that of their children.

- Using the table of human well-being rankings (Table 8.1), shade in each country's female literacy rate from light to dark using these categories: 0-30 percent; 31-50 percent; and 51-100 percent.
- Referring to Figure 8.15 (declines in total fertility rates), use a graduated symbol (from small to large circles, for example) to map the 2005 total fertility rate for the countries provided, using these categories: 0-2; 2.1-4; and 4.1-5.
- Finally, write in the number for the Gender Development Index (GDI) (Table 8.1).

Questions

a. Analyzing the information you mapped, describe what spatial pattern(s) you see.

b. Does population growth (total fertility rate) generally increase or decrease with education (female literacy rate)?

c. Discuss how GDI relates to education (literacy rate) and population growth (total fertility rate).

d. Do you see any anomalies (e.g., high literacy/high growth rate or low literacy/low growth rate)? Why might this be the case?

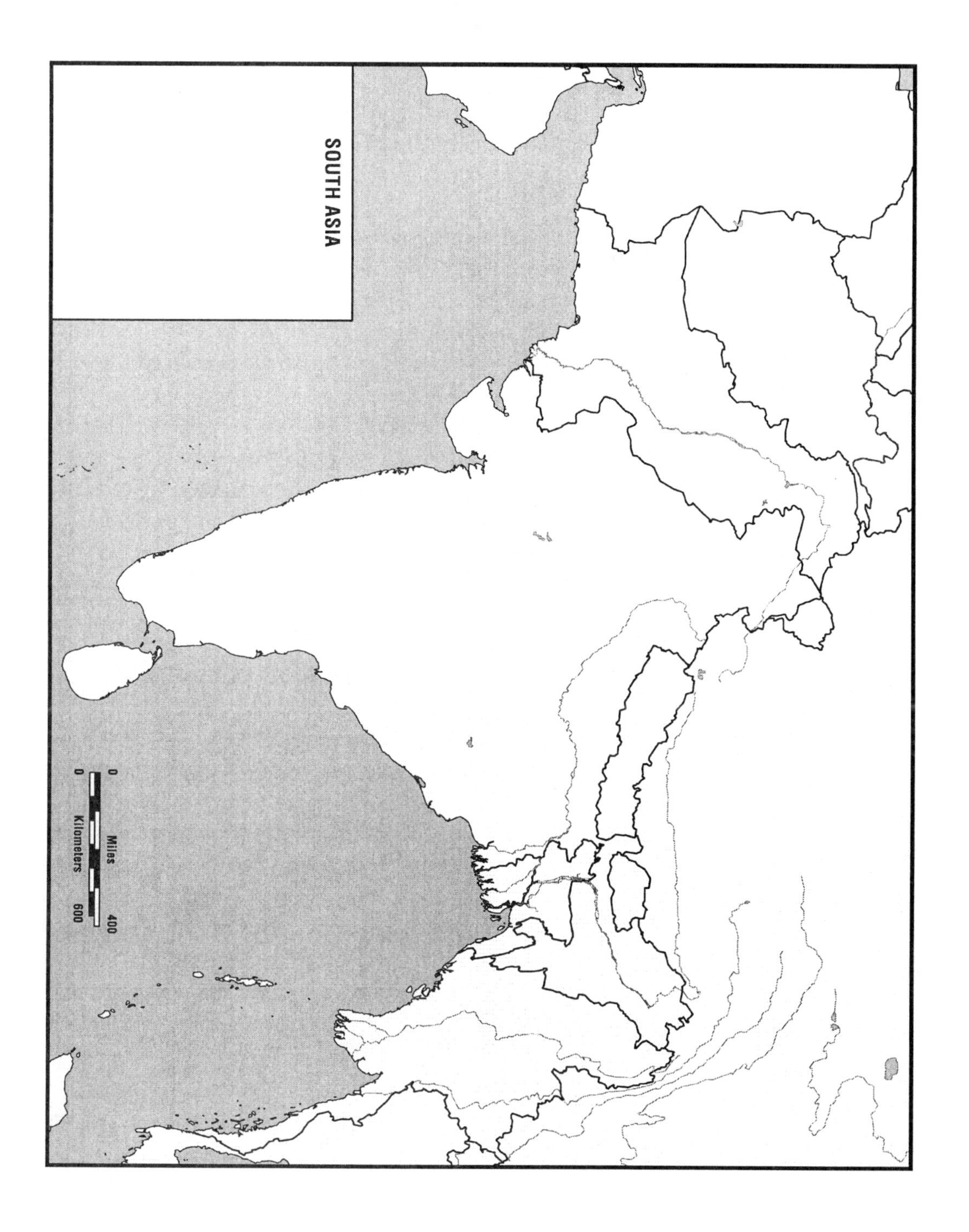
SOUTH ASIA
0
Miles
400
0
Kilometers
600

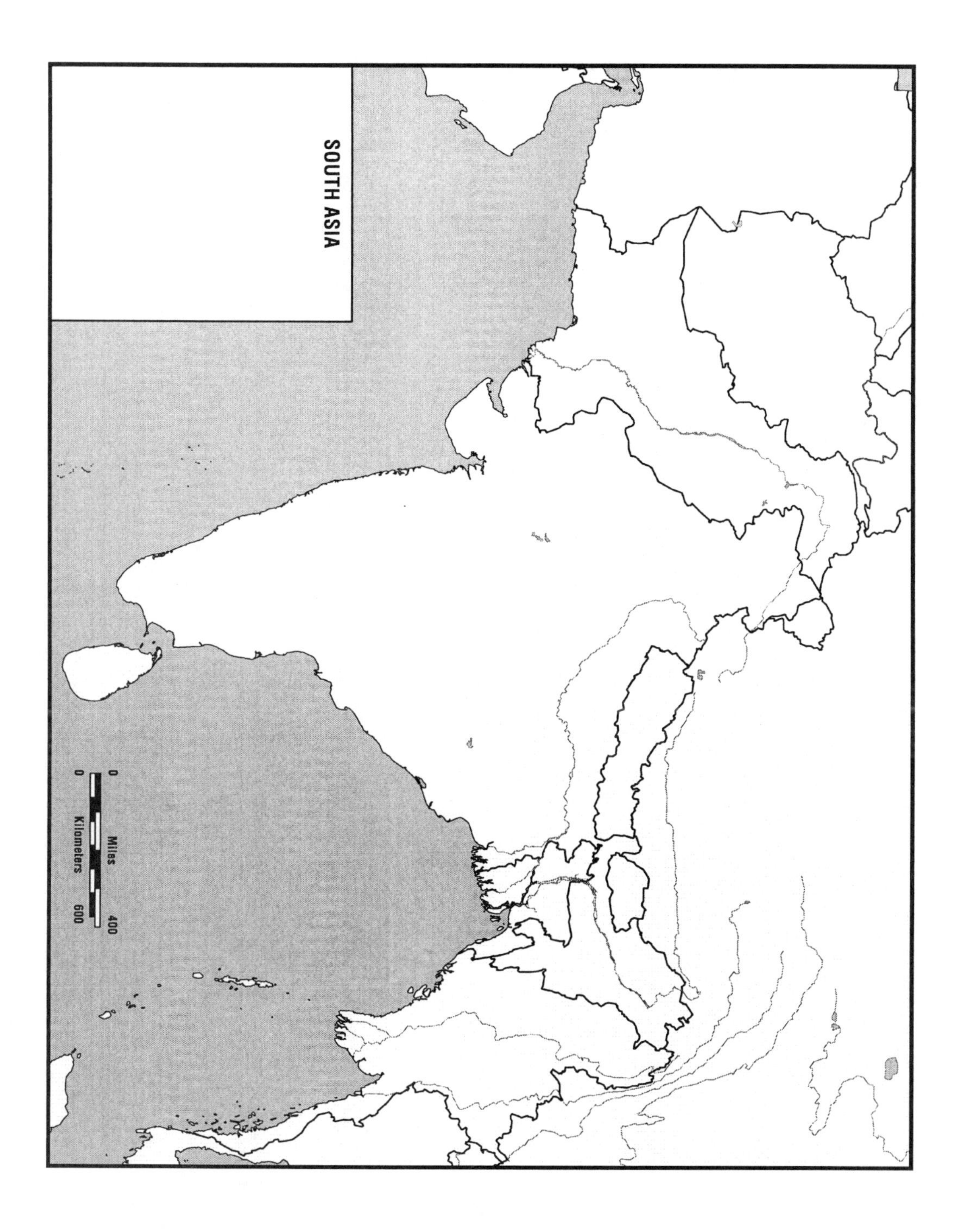
SOUTH ASIA
0
Miles
400
0
Kilometers
600

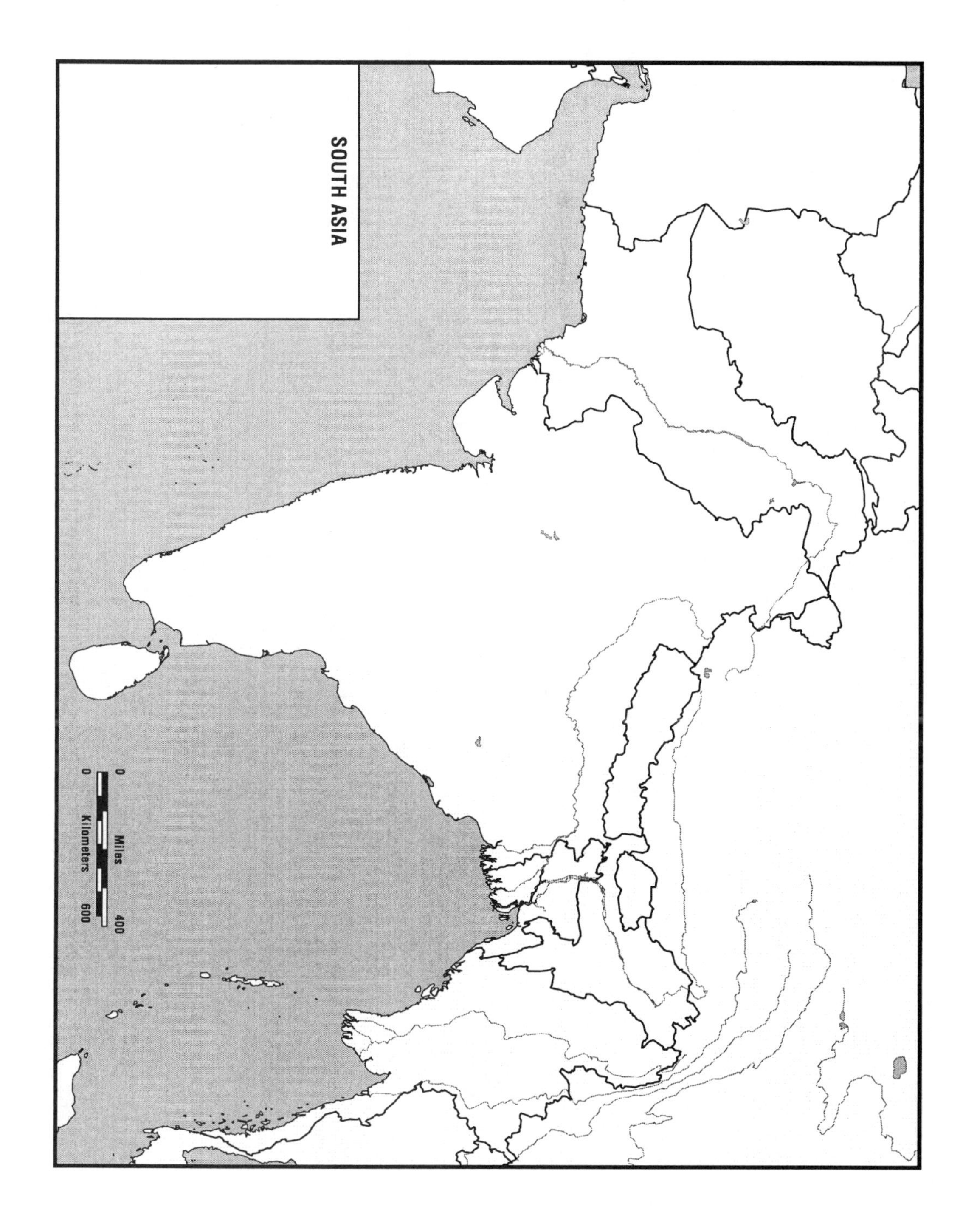
SOUTH ASIA
0 Miles 400
0 Kilometers 600

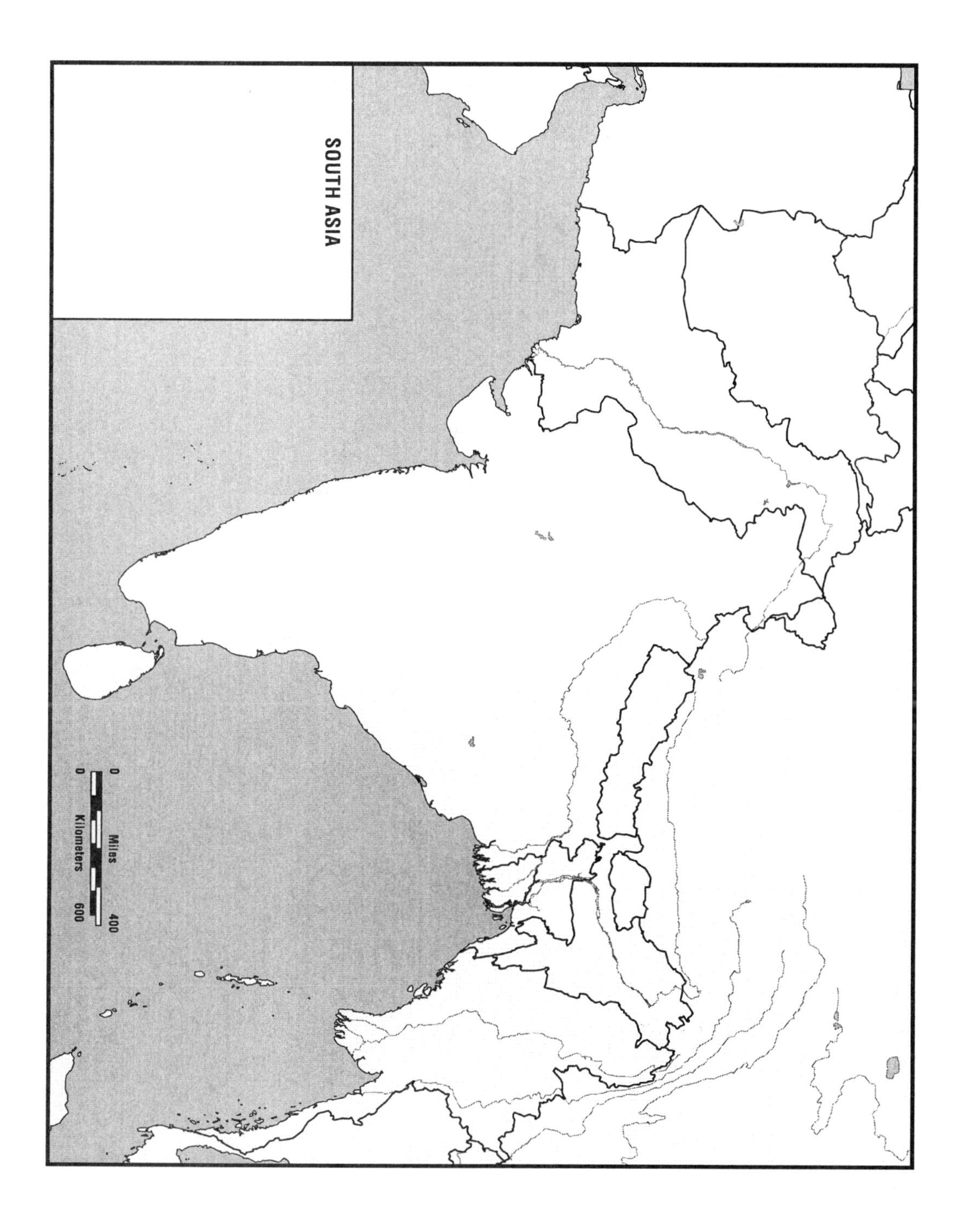
SOUTH ASIA
Miles
0
400
Kilometers
0
600

CHAPTER NINE
East Asia

IMPORTANT PLACES

The following places are featured in the chapter. Make sure you can locate all of them on a map. A blank outline map of the region is provided at the end of this chapter; additional maps can be found on the textbook's Web site: www.whfreeman.com/pulsipher4e. Also, to prepare for exams, write a few important facts about each place in the space provided.

Physical Features

1. Chang Jiang (Yangtze)

2. East China Sea

3. Gobi Desert

4. Himalayas

5. Huang He (Yellow River)

6. Junggar Basin

7. Mekong River

8. Mongolian Plateau

9. Mount Fuji

10. North China Plain

11. Nu (Salween River)

12. Ordos Desert

13. Pacific Ocean

14. Pacific Ring of Fire

15. Plateau of Tibet (Xizang-Qinghai Plateau or Northern Plateau)

16. Qaidam Basin

17. Sea of Japan

18. Sichuan Basin

19. South China Sea

20. Taklimakan Desert

21. Tarim Basin

22. Zhu Jiang (Pearl River)

23. Yellow Sea

24. Yunnan Guizhou Plateau

Regions/Countries/States/Provinces

25. China

26. China's Far Northeast

27. Guangdong

28. Hong Kong

29. Inner Mongolia (Nei Monggol)

30. Japan

31. Macau

32. Mongolia

33. North Korea

34. South Korea

35. Taiwan

36. Tibet (Xizang)

37. Xinjiang Uygur

Cities/Urban Areas

38. Beijing

39. Guangzhou

40. Kyoto

41. Osaka

42. Pyongyang

43. Seoul

44. Shanghai

45. Taipei

46. Tokyo

47. Ulan Bator

MAPPING EXERCISES

The following are three mapping exercises to improve your knowledge of the location of places, underscore why they are important, and clarify how they relate to each other. Some questions will ask you to locate places, compare maps, or fill in data; others will test your understanding of *why* you were asked to map the features that you did. Use the blank outline maps at the end of the chapter to complete these exercises. Additional blank outline maps can be found on the textbook's Web site: www.whfreeman.com/pulsipher4e.

1. Population and landforms

Of the few flat surfaces in the rugged landscapes of East Asia, many are too cold or dry to be useful to humans.

- On a blank map of East Asia, shade the areas (in red) that have over 1300 persons per square mile (over 500 people per square kilometer) (Figure 9.14).
- From Figure 9.1 (regional map), shade (in tan) and label the location of the Gobi, Ordos, and Taklimakan deserts.
- Trace with a thick blue line and label the following rivers: Chang Jiang (Yangtze), Huang He (Yellow River), Mekong River, Nu (Salween), and Zhu Jiang (Pearl River).
- Draw (with ^^^) and label the location of the Himalayas.
- Draw a hatch pattern over and label the Plateau of Tibet and Yunnan Guizhou Plateau.

Questions

a. Is population concentrated near any particular landform(s)?
b. Which landforms are associated with low/no population concentrations?
c. In cases where population is not located near a water source, how might people adapt to make agriculture productive?
d. In cases where population is located near rugged terrain, what adaptations might people make to create space for agriculture?

2. Economic growth, SEZs, and ETDZs

Special economic zones (SEZs) and economic and technology development zones (ETDZs) are central to China's new market reforms and have rapidly opened the economy to international trade.

- Using Figure 9.22 (map of foreign investment), on the blank map of China and its provinces, draw a blue square for each SEZ and a red dot for each ETDZ.
- Using Figure 9.21 (map of rural-urban GDP disparities), shade the provinces in green that have over 9000 yuan GDP per capita (use yellow for the "upper middle" category and orange for the "high" category).

Questions

a. What type of general relationship did you expect to find between the SEZs/ETDZs and GDP per capita? Explain why SEZs/ETDZs might affect GDP per capita.
b. Are there provinces with SEZs/ETDZs that don't have an upper-middle or high GDP? Provide at least two reasons why do you think this is the case.
c. Assume that SEZs/ETDZs are major growth poles (drawing more investment and migration) and examine the map of population density (Figure 9.14) and the map of rural-to-urban migration (Figure 9.3).
    - What will happen to those provinces with SEZs/ETDZs that already have high population densities (consider the environment, infrastructure, and resulting living conditions)?
    - Why do you think the government has chosen to put SEZs/ETDZs in areas with population densities? What effects might SEZs/ETDZs have on these areas?

3. The Three Gorges Dam

Although the Three Gorges Dam is expected to save millions of lives and much property, the 370-mile long reservoir it creates will drown 62,000 acres of farmland, 13 major cities, 140 large towns, hundreds of small villages, and 1600 factories, displacing over 1.9 million people.

- On the blank map of the Central China subregion, draw the Chang Jiang with a thick blue line.
- Using the map of population density (Figure 9.14), shade in areas that have over 1300 people per square mile (over 500 people per square kilometer).
- Using the map of foreign investment (Figure 9.22), draw a red circle for each ETDZ and a blue square for each SEZ.

- Using the maps on the University of Hong Kong's Civil Engineering Computer Aided Learning (CIVCAL) Web site (civcal.media.hku.hk/threegorges/Default.htm), draw in the location of the dam.
- Also, using the maps on the CIVCAL Web site, draw the outline, showing the extent of the reservoir that will be created by the Three Gorges Dam.

Questions

a. Based on the outline of the reservoir, what cities or urban areas might be affected and how?
b. Will this dam negatively affect any of the ETDZs or SEZs by flooding them or reducing the amount of water flowing in the area? Which ones?
c. Will the dam/reservoir bring any ETDZs or SEZs *more* business? Which ones and why?
d. Using the University of Hong Kong's Web site, identify two positive effects the dam/reservoir might have on each of the following: society, politics, and the environment.
e. Evaluate the University of Hong Kong's Web site: What type of information is missing from this site? Find at least two other critical Web sites and identify at least five negative effects the dam/reservoir might have on: society, politics, and the environment.

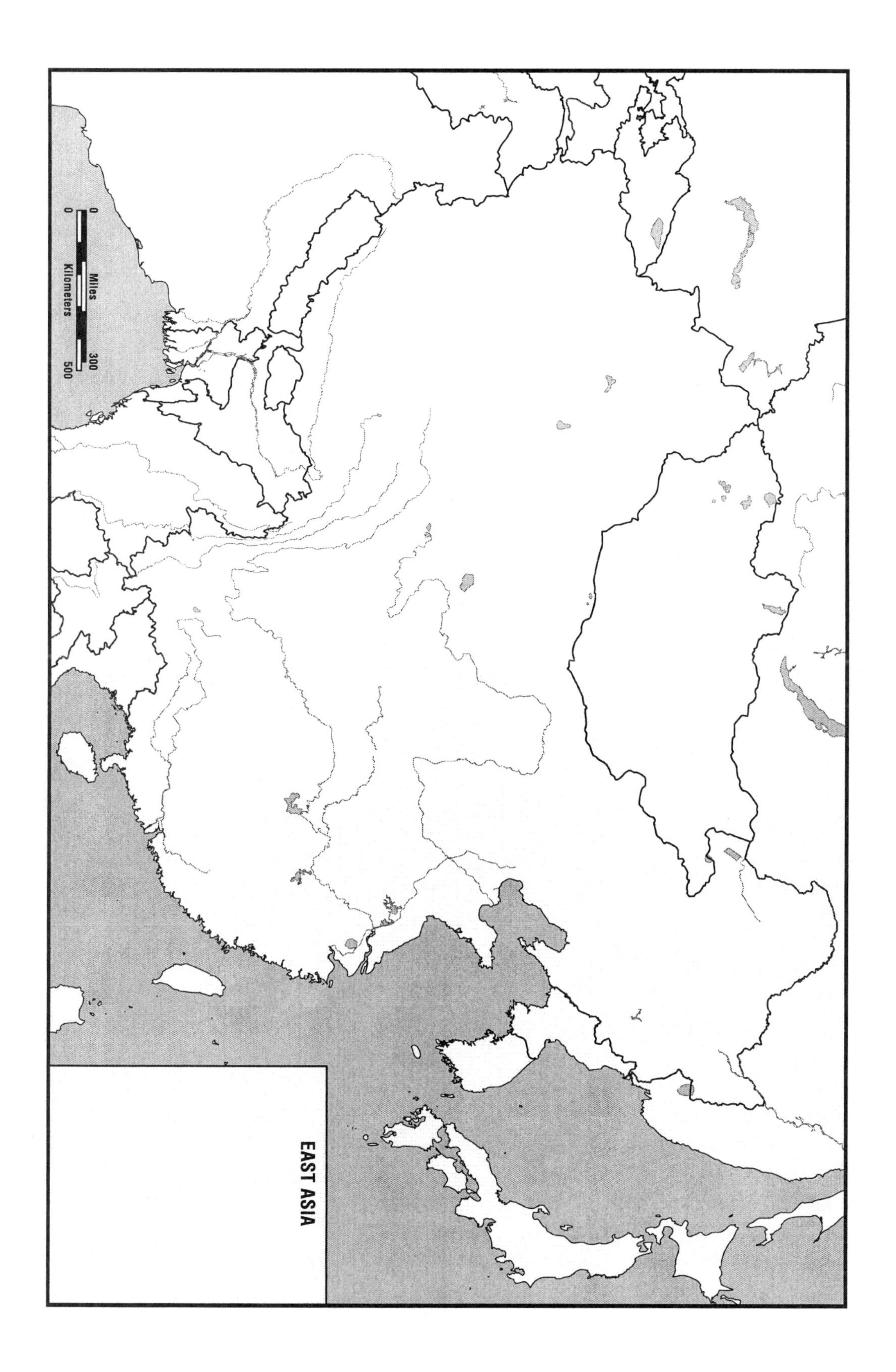

EAST ASIA
0
Miles
300
0
Kilometers
500

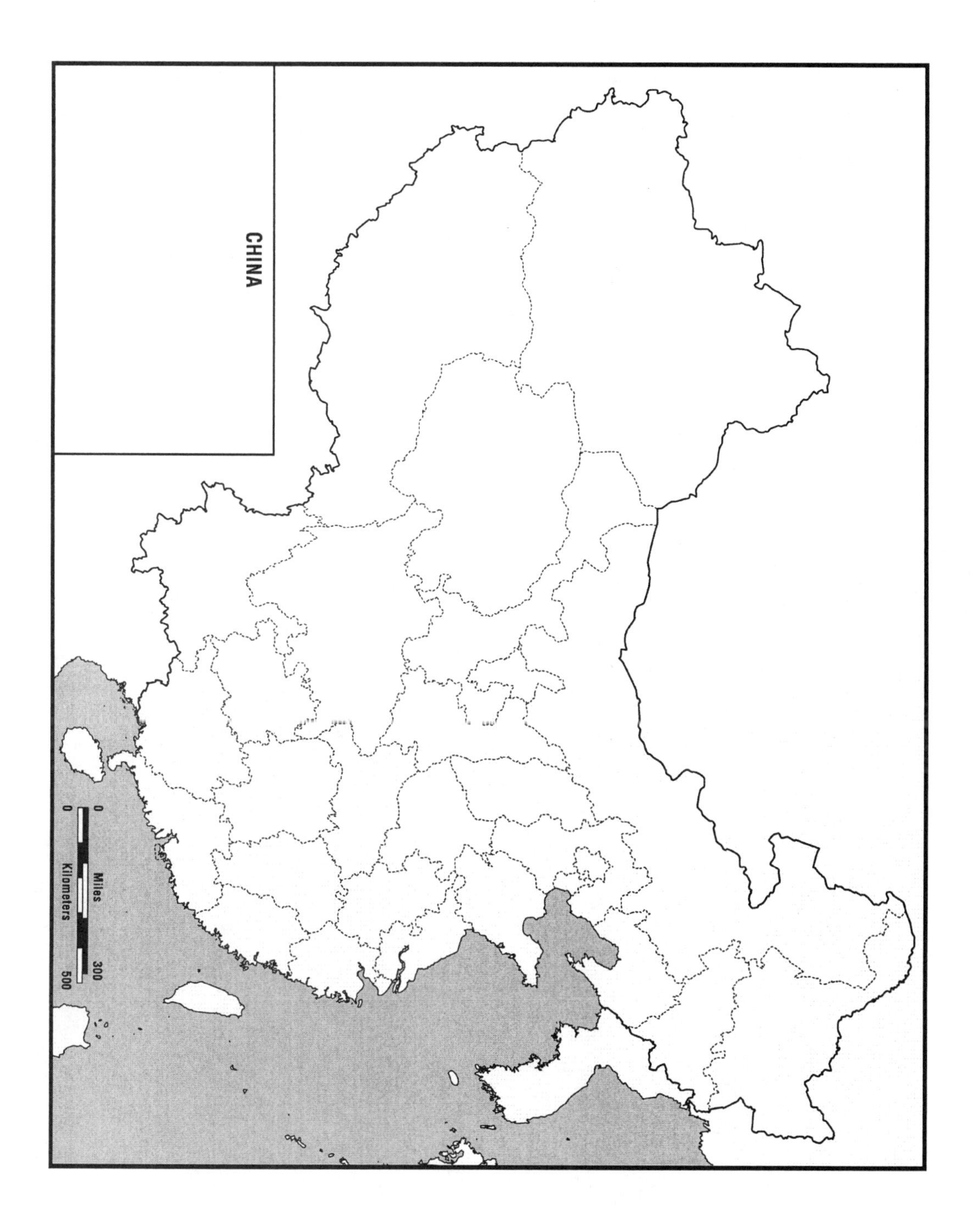
CHINA
0
Miles
300
0
Kilometers
500

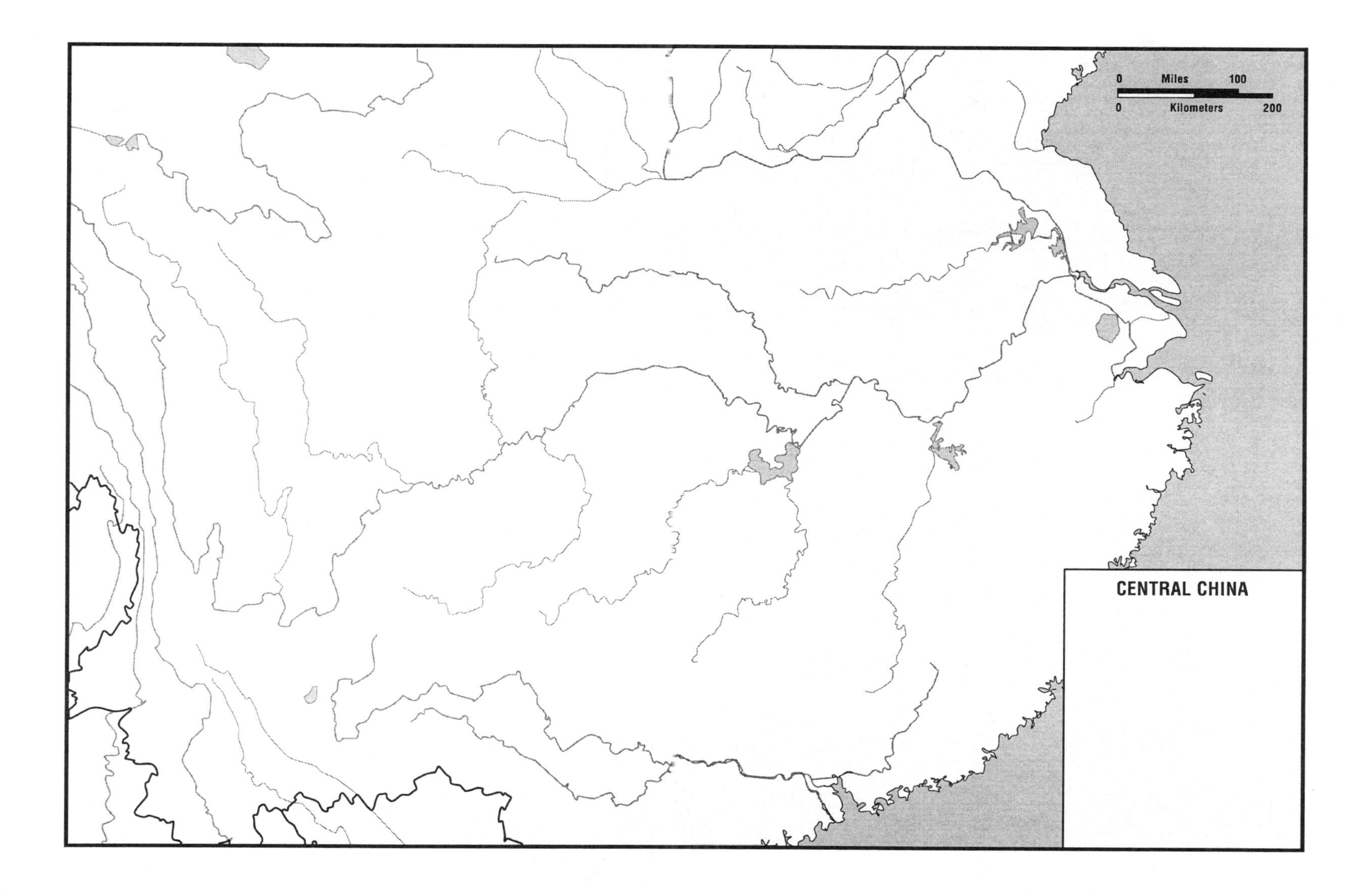
0
Miles
100
0
Kilometers
200
CENTRAL CHINA

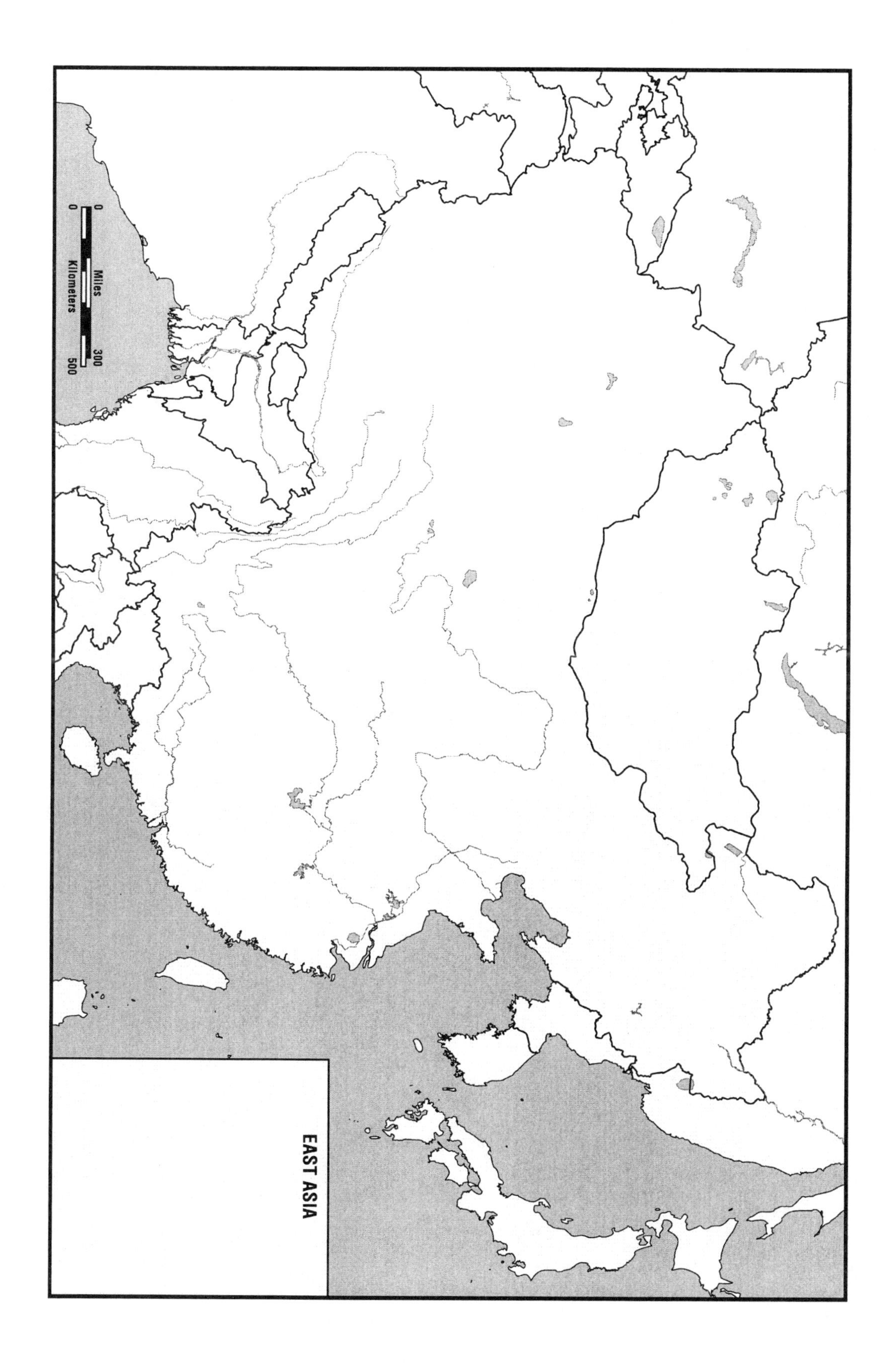
EAST ASIA
0 Miles 300
0 Kilometers 500

CHAPTER TEN
Southeast Asia

IMPORTANT PLACES

The following places are featured in the chapter. Make sure you can locate all of them on a map. A blank outline map of the region is provided at the end of this chapter; additional maps can be found on the textbook's Web site: www.whfreeman.com/pulsipher4e. Also, to prepare for exams, write a few important facts about each place in the space provided.

Physical Features

1. Bali
2. Black River
3. Borneo
4. Chao Phraya River
5. Gulf of Thailand
6. Irrawaddy River
7. Java
8. Lesser Sunda Islands
9. Luzon
10. Madura
11. Malay Peninsula
12. Mekong River
13. Mindanao
14. Molucca Islands
15. Mount Pinatubo
16. New Guinea

17. Red River

18. Salween River

19. South China Sea

20. Strait of Malacca

21. Sulawesi (Celebes Islands)

22. Sumatra

23. Timor Island

Regions/Countries/States/Provinces

24. Brunei

25. Burma (Myanmar)

26. Cambodia

27. East Timor (Timor-Leste)

28. Indochina

29. Indonesia

30. Kalimantan

31. Laos

32. Malaysia

33. Philippines

34. Sabah

35. Sarawak

36. Singapore

37. Thailand

38. Vietnam

39. West Papua

Cities/Urban Areas

40. Bandar Seri Begawan

41. Bandung

42. Bangkok

43. Chiang Mai

44. Hanoi

45. Ho Chi Minh City (Saigon)

46. Jakarta

47. Kuala Lumpur

48. Manila

49. Phnom Penh

50. Rangoon

51. Vientiane

MAPPING EXERCISES

The following are three mapping exercises to improve your knowledge of the location of places, underscore why they are important, and clarify how they relate to each other. Some questions will ask you to locate places, compare maps, or fill in data; others will test your understanding of *why* you were asked to map the features that you did. Use the blank outline maps at the end of the chapter to complete these exercises. Additional blank outline maps can be found on the textbook's Web site: www.whfreeman.com/pulsipher4e.

1. Tourism and indigenous peoples

Tourism is suggested as a good alternative for regional and local economic development in Southeast Asia. However, indigenous peoples may be adversely affected while they, at the same time, reap economic benefits.

- Using Figure 10.3 (map of indigenous groups), and especially the inset map, shade the indigenous groups' locations on a blank map of Southeast Asia.
- Draw in the approximate location of the Asian highway system, using Figure 10.20 (map of transportation infrastructure).
- Also using Figure 10.20, mark the location of world heritage sites with a black dot.
- Finally, use Figure 10.10 (population density) to cross hatch (///) areas with densities of 651 or more people per square mile (251 or more people per square kilometer).

Questions

a. Based on your map, as well as the reading of the first few pages of the chapter, the section entitled "Tourism Development," and Box 10.1 (Tourism Development in the Greater Mekong Basin), identify those areas where indigenous peoples will be highly affected by a developing tourist industry.
b. Suggest benefits to the indigenous peoples in the places you have identified. Make a judgment as to the sustainability of those benefits.
c. Suggest special challenges these indigenous people may face and discuss how their culture can be protected.

2. Agriculture and population density

Population density can often be related to economic lifestyles; even varying types of agricultural systems are often associated with different population densities.

- On a blank map of Southeast Asia, label all of the countries in the region.
- Using Figure 10.16 (map of agricultural patterns), shade intensive cropland.
- Using Figure 10.10 (map of population density), cross hatch (///) the areas with 651 or more people per square mile (251 or more people per square kilometer).

Questions

a. What is the relationship between intensity of agriculture and population density?
b. Are there any anomalies? If so, where are they? Explain them.

3. The diaspora of Southeast Asian women

Over 50 percent of the people who engage in extra regional migration are women. Many of these women move within the region as well as to other world regions to find work as maids.

- Using Table 10.4 (human well-being rankings), draw a graduated circle in each country to represent its GDP. Remember to make the smallest circle correspond to the lowest range of GDP.
- Again, using Table 10.4, select gray shades to shade female literacy for all countries. Remember to use the lightest gray for the lowest levels of literacy and the darkest shade for the highest level of literacy.

Questions

a. Refer to the "maid trade" map (Figure 10.30) and identify three countries in this region that appear to be major contributors to the "maid trade." Explain any relations you see between GDP and female literacy and countries' contribution to the "maid trade."
b. If no relationship is suggested by the map, then using your understanding of this region, suggest other reasons for a country's high "maid trade" numbers.

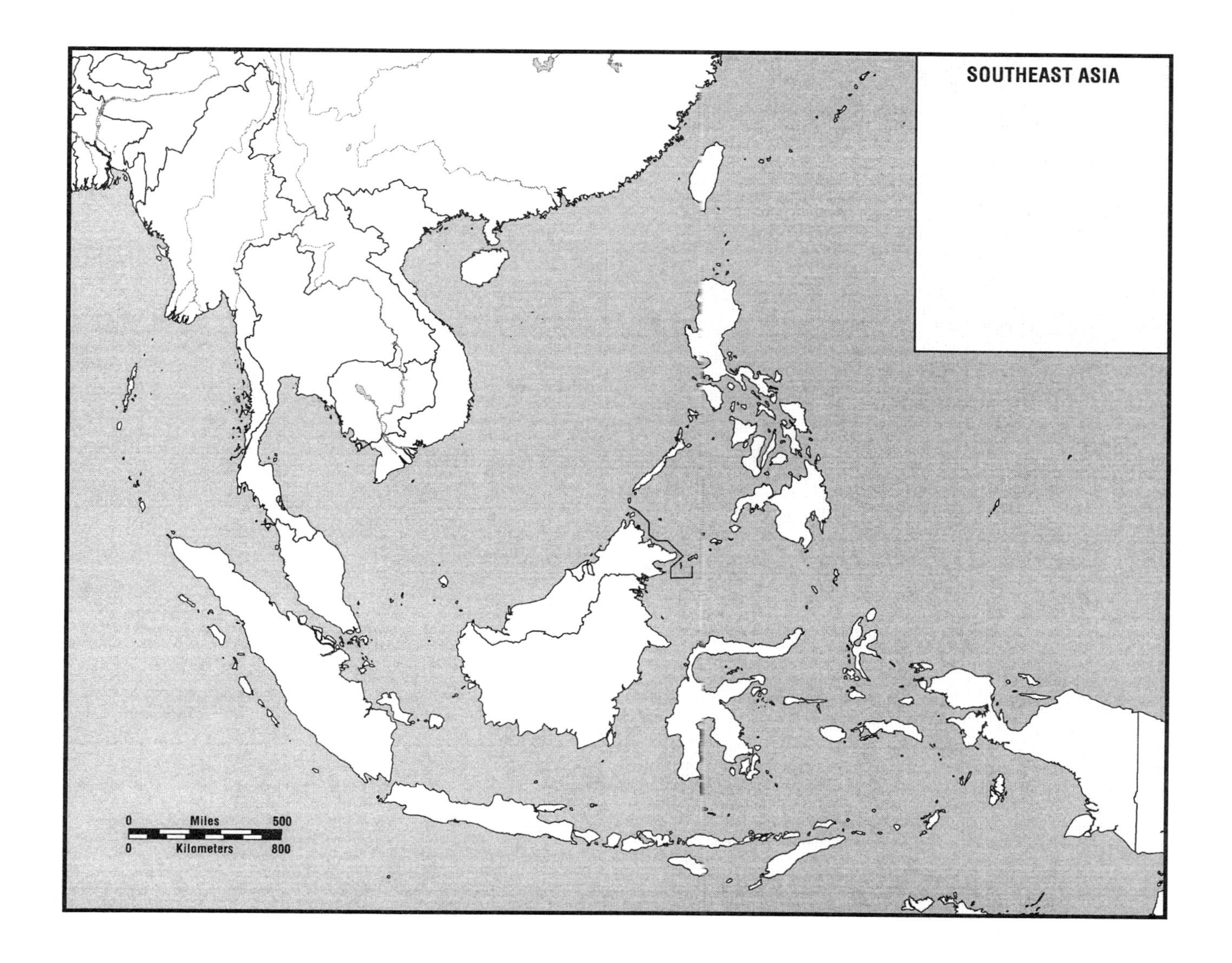
SOUTHEAST ASIA
0 Miles 500
0 Kilometers 800

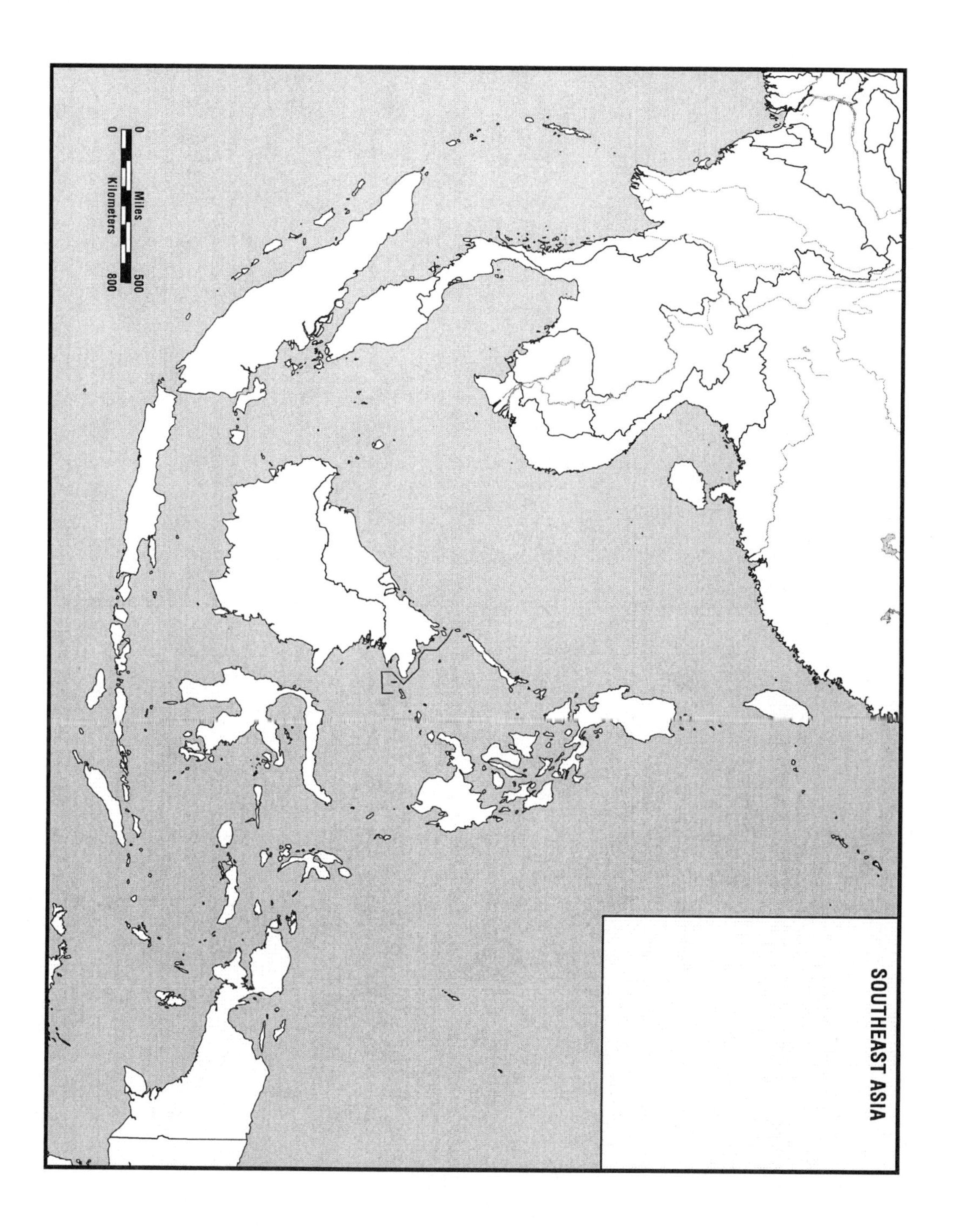
SOUTHEAST ASIA
Miles
0 500
Kilometers
0 800

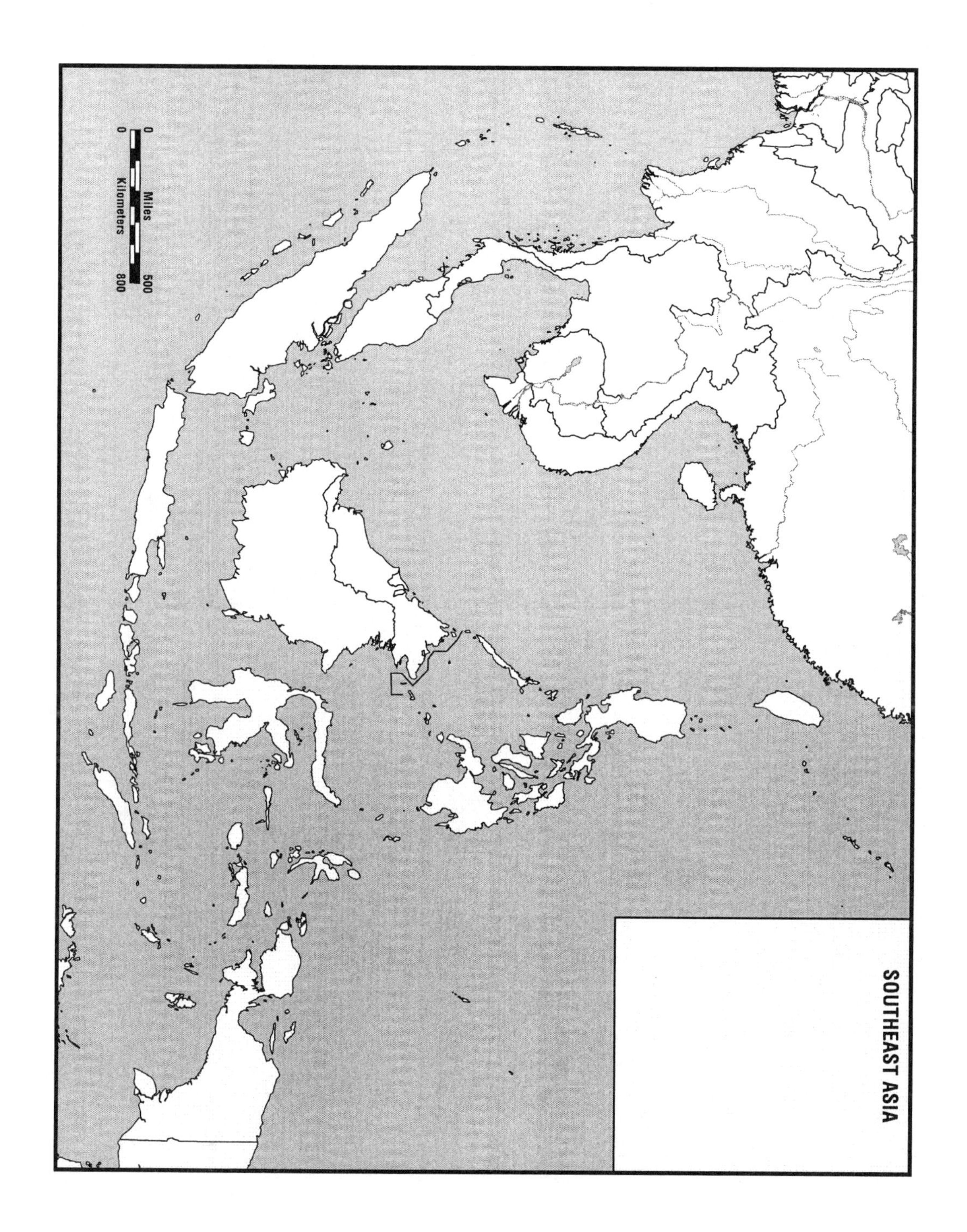
0 Miles 500
0 Kilometers 800
SOUTHEAST ASIA

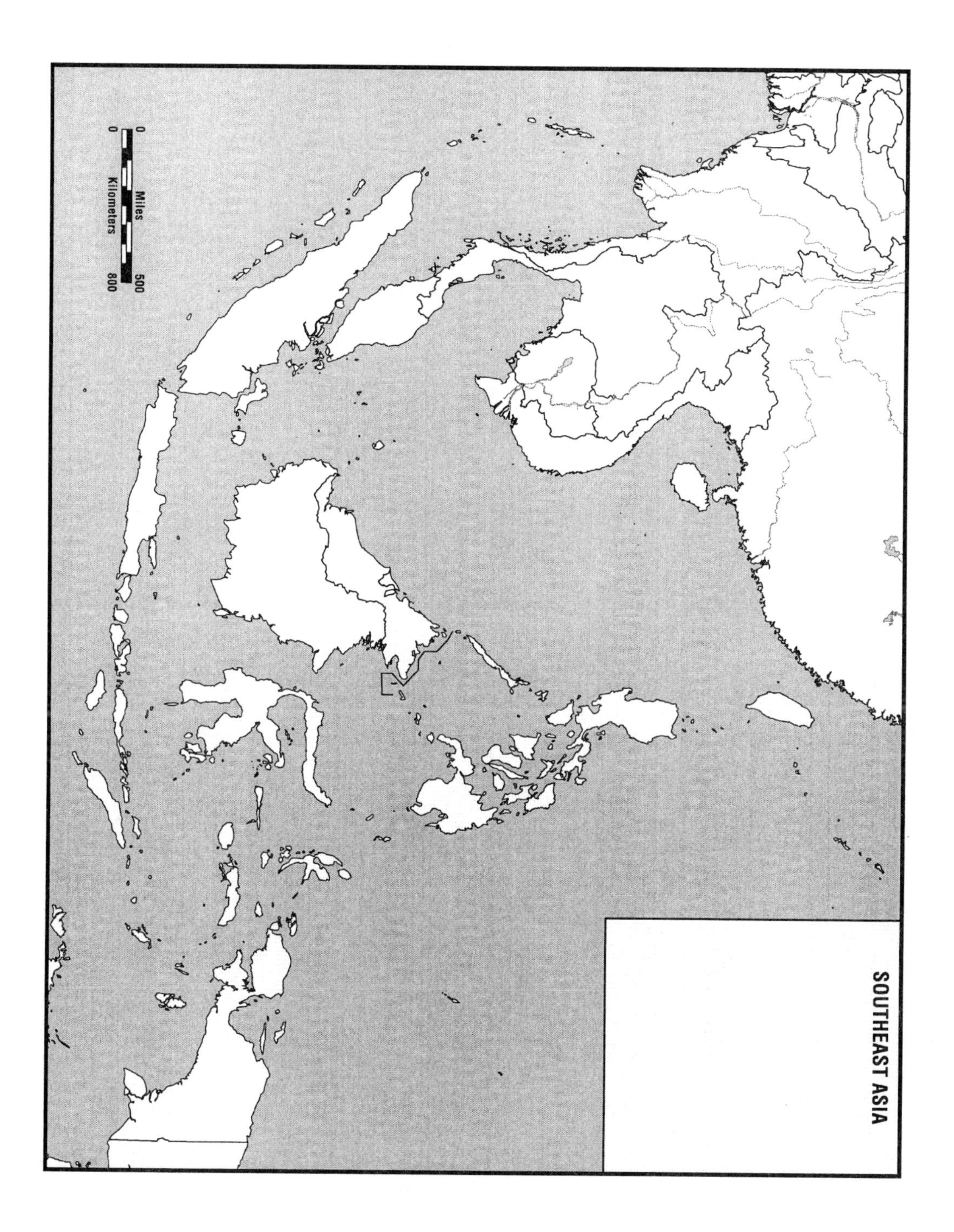

SOUTHEAST ASIA
0
500
Miles
0
800
Kilometers

CHAPTER ELEVEN

Oceania: Australia, New Zealand, and the Pacific

IMPORTANT PLACES

The following places are featured in the chapter. Make sure you can locate all of them on a map. A blank outline map of the region is provided at the end of this chapter; additional maps can be found on the textbook's Web site: www.whfreeman.com/pulsipher4e. Also, to prepare for exams, write a few important facts about each place in the space provided.

Physical Features

1. Ayers Rock (Uluru)

2. Caroline Islands

3. Cook Islands

4. Coral Sea

5. Darling River

6. Easter Island

7. Eastern Highlands (Great Dividing Range)

8. Galápagos Islands

9. Great Australian Bight

10. Great Barrier Reef

11. Gulf of Carpentaria

12. Indian Ocean

13. Murray River

14. New Guinea

15. Pacific Ocean

16. Samoa Islands

17. Southern Alps

18. Tasman Sea

19. Tasmania

Regions/Countries/States/Provinces

20. American Samoa

21. Australia

22. Federated States of Micronesia

23. Fiji

24. French Polynesia

25. Guam

26. Hawaii

27. Kiribati

28. Marshall Islands

29. Melanesia

30. Micronesia

31. Nauru

32. New Caledonia

33. New Zealand

34. Northern Marianas

35. Palau

36. Papua New Guinea

37. Polynesia

38. Samoa (Western Samoa)

39. Solomon Islands

40. Tahiti

41. Tonga

42. Tuvalu

43. Vanuatu

Cities/Urban Areas

44. Adelaide

45. Apia

46. Auckland

47. Brisbane

48. Canberra

49. Christchurch

50. Funafuti

51. Honiara

52. Honolulu

53. Koror

54. Majuro

55. Melbourne

56. Newcastle

57. Nuku'alofa

58. Palikir

59. Papeete

60. Perth

61. Port Arthur

62. Port Moresby

63. Port Vila

64. Suva

65. Sydney

66. Tarawa

67. Wellington

68. Yaren

MAPPING EXERCISES

The following are three mapping exercises to improve your knowledge of the location of places, underscore why they are important, and clarify how they relate to each other. Some questions will ask you to locate places, compare maps, or fill in data; others will test your understanding of *why* you were asked to map the features that you did. Use the blank outline maps at the end of the chapter to complete these exercises. Additional blank outline maps can be found on the textbook's Web site: www.whfreeman.com/pulsipher4e.

1. Physical geography and population

The physical geography of Australia may play an important role in population distribution.

- Using Figure 11.1 (regional map), select five specific physical features (e.g., mountain ranges, coastal zones, and deserts) for Australia and delineate them on the blank map of Australia. Include an appropriate legend. Using Figure 11.8 (climate map), select three climatic zones and depict them on your map. Include an appropriate legend.
- Draw a dot and label cities with populations of 1 million or greater, as depicted in Figure 11.16 (population density map).

Questions

a. Justify your choice of physical features. Justify your choice of climatic zones.
b. Make at least two generalizations about Australia's population concentrations based on your selection of physical geography characteristics.
c. Make at least two generalizations about Australia's population concentrations based on your selection of climatic zones.
d. What are some unique challenges faced by people inhabiting regions with low population densities?
e. What are some challenges faced by people inhabiting regions with high population densities?

2. The United Nations Convention on the Law of the Sea and Oceania

The UN Convention on the Law of the Sea has particular impacts on this region.

- On a blank map of Oceania, approximate the 200-mile exclusive economic zone that each country can exploit. Draw this border around the island nations. You will have to make generalizations because of the scale at which you will be working.

Questions

a. The United Nations Convention on the Law of the Sea designates that countries sharing overlapping areas will draw a border midway between their coasts. Use a cross hatch marking (///) to identify any areas that may be in conflict. Suggest three possible conflicts that may arise.
b. Based on your understanding of Oceania, suggest benefits exclusive to these nations that are a result of the Law of the Sea Treaty.
c. Who do you think are the major global participants in exploiting the Exclusive Economic Zones? How can small independent island nations be assertive in defining cooperation with those who want to "lease" these zones?
d. Compare the gains and losses discussed above and decide the overall value of the Law of the Sea Treaty to Oceania.

3. Global warming threatens Oceania

A single meter rise in the sea level can have profound impacts on some of the islands and countries discussed in this chapter.

- Using a blank map of Oceania, color-code the islands according to their status of high island or low island. Refer to the CIA World Factbook at www.cia.gov/cia/publications/factbook for information about elevation minimums.

Questions

a. Suggest at least two impacts that both types of islands will experience if global warming results in a one meter rise in ocean levels. Identify at least two additional impacts that only the low islands will experience. In all cases, justify your reasons for selecting these particular impacts.
b. Identify strategies that these islands should follow to prevent catastrophic effects from global warming. What strategies are specific to high islands? What strategies are specific to low islands?

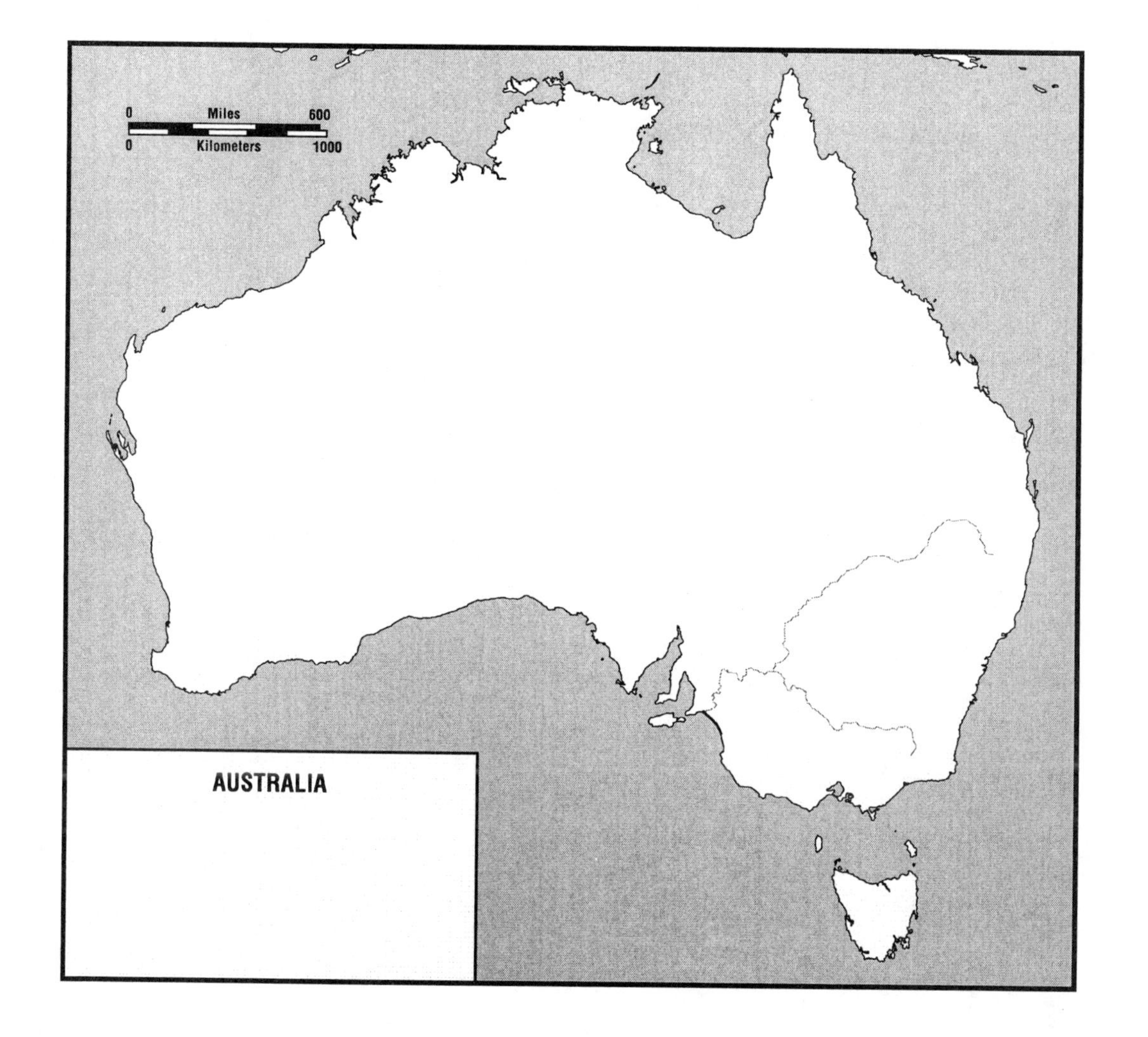
0 Miles 600
0 Kilometers 1000
AUSTRALIA

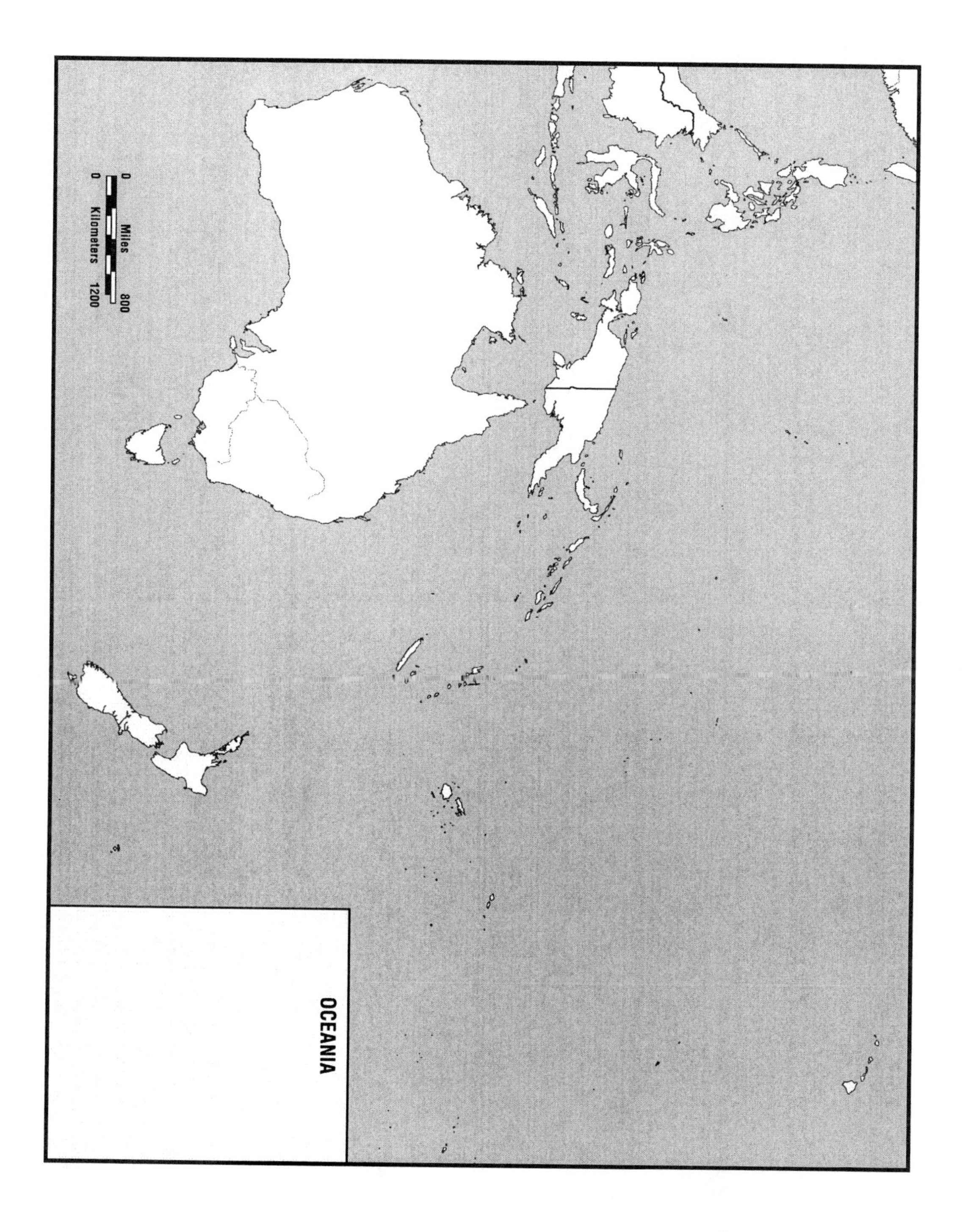
OCEANIA
0 Miles 800
0 Kilometers 1200

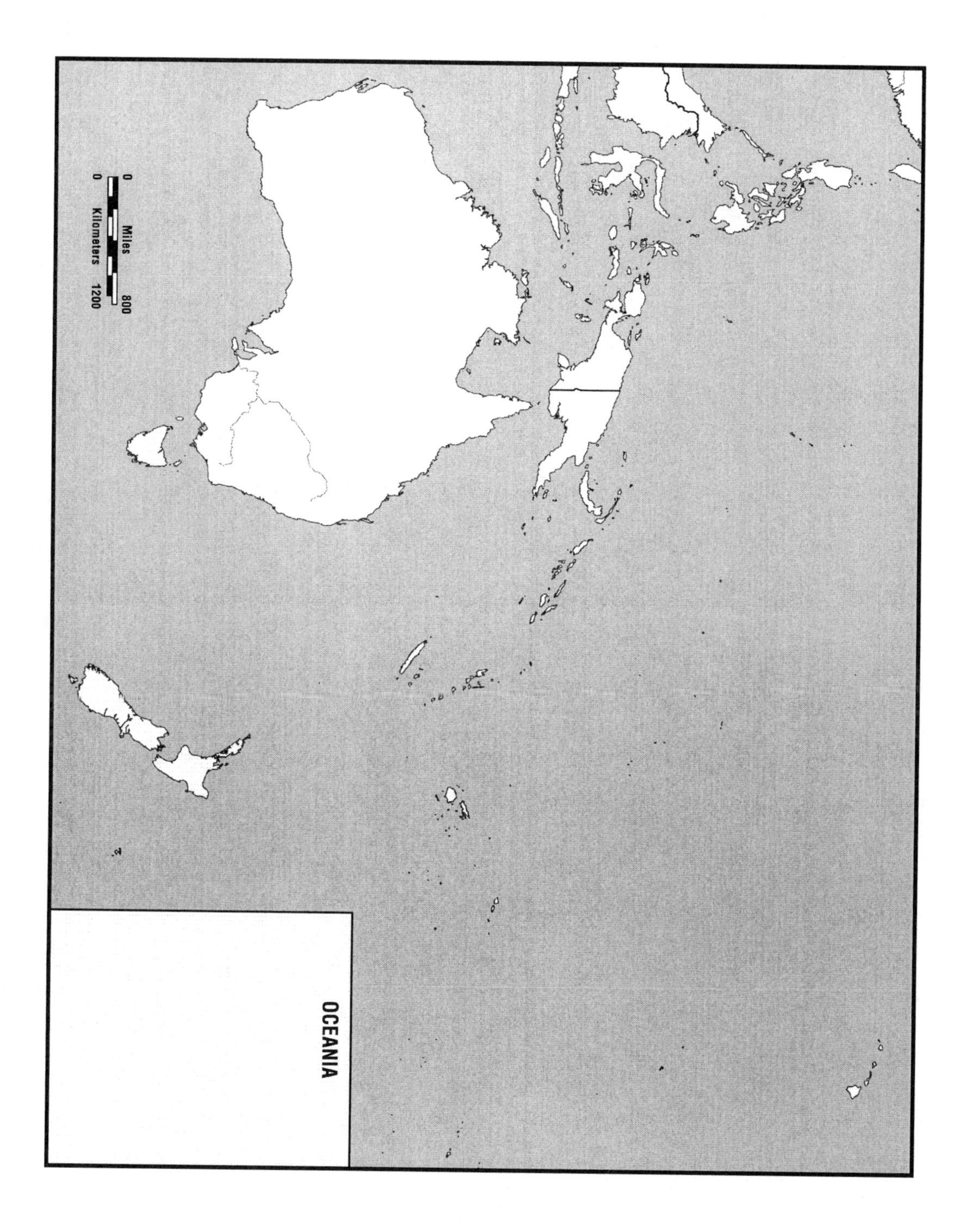

OCEANIA
Miles
0
800
Kilometers
0
1200

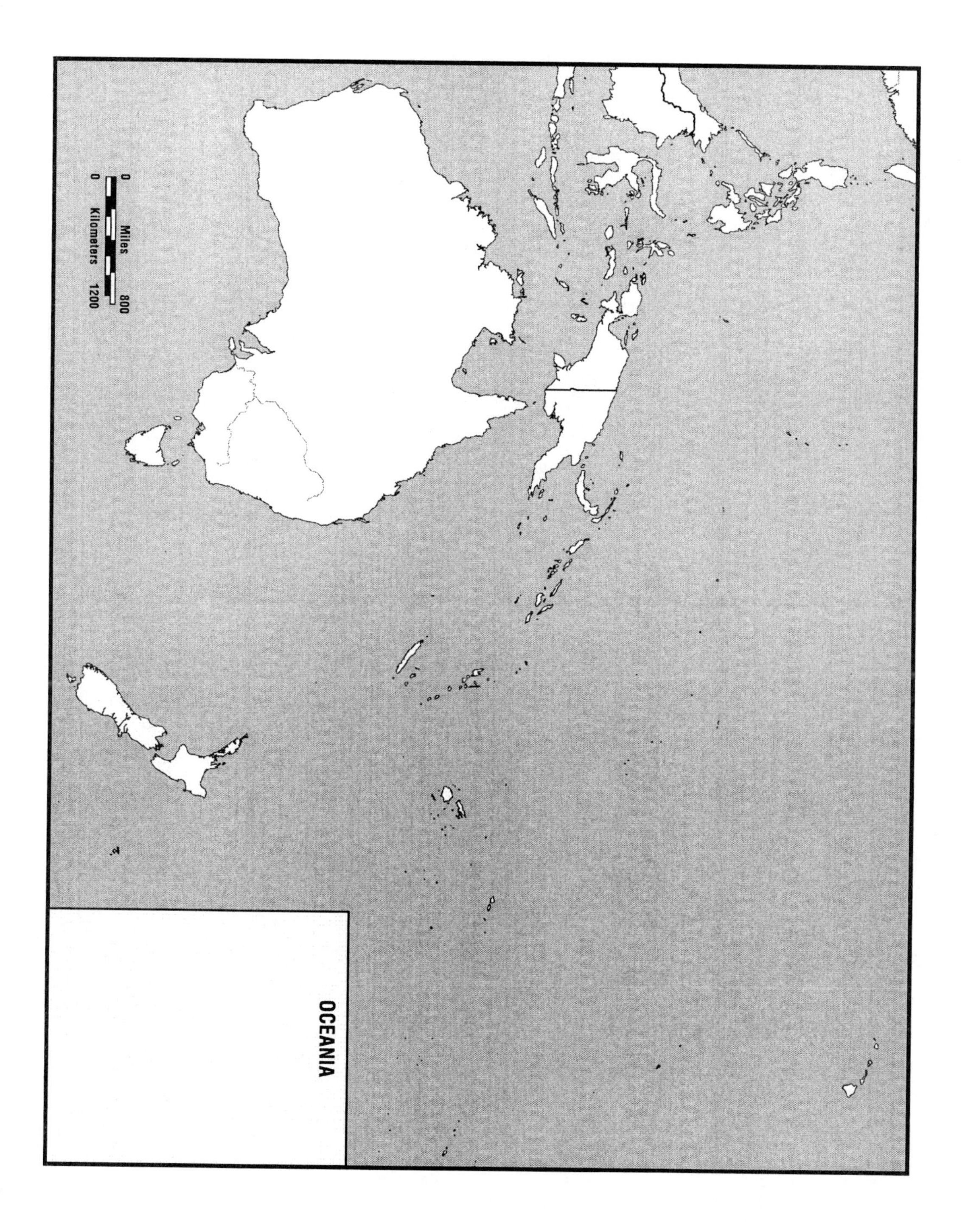
OCEANIA
0 Miles 800
0 Kilometers 1200

APPENDIX

BLANK WORLD MAPS

Additional blank maps can be found at: www.whfreeman.com/pulsipher4e

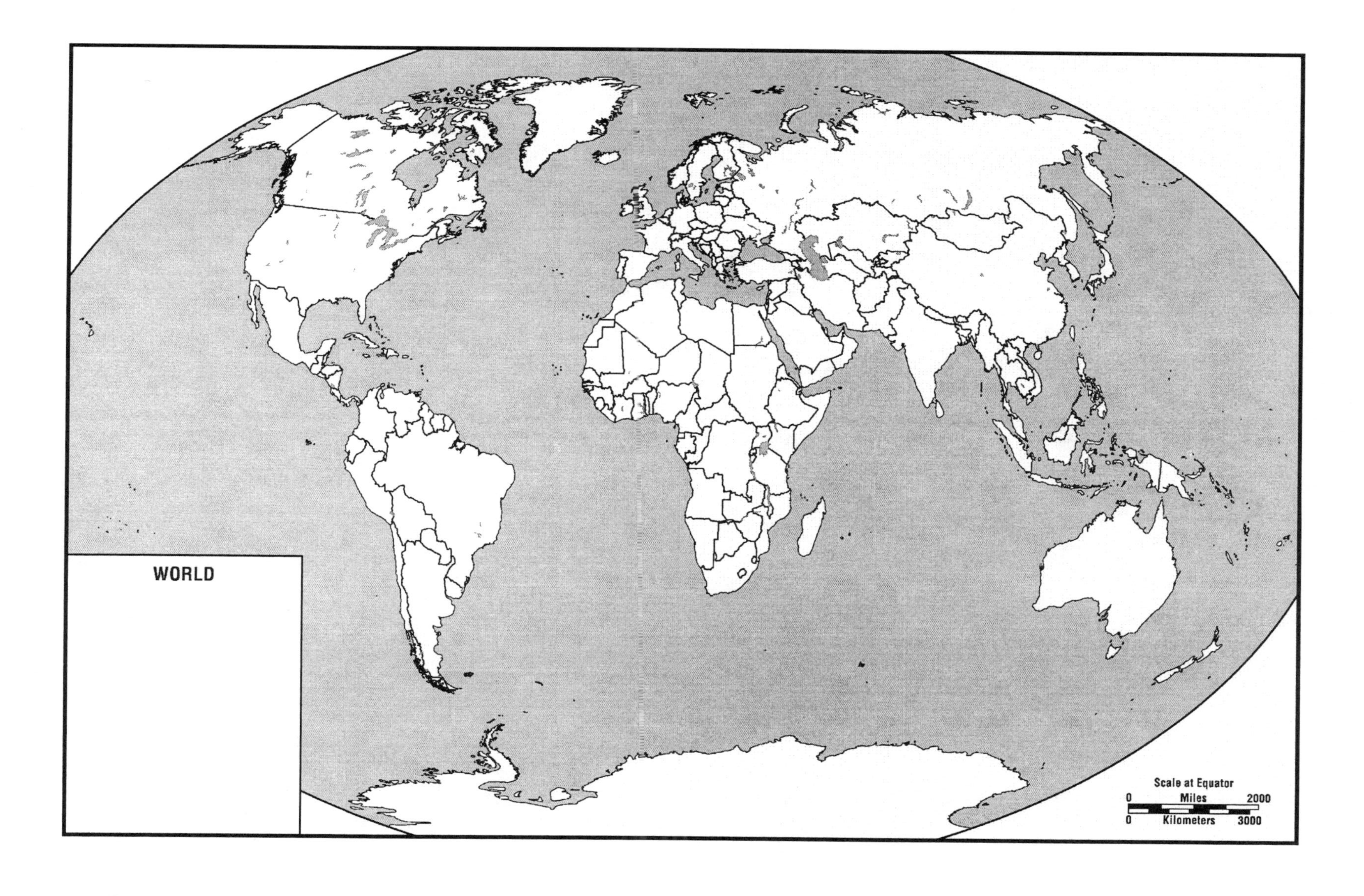
WORLD
Scale at Equator
0 Miles 2000
0 Kilometers 3000

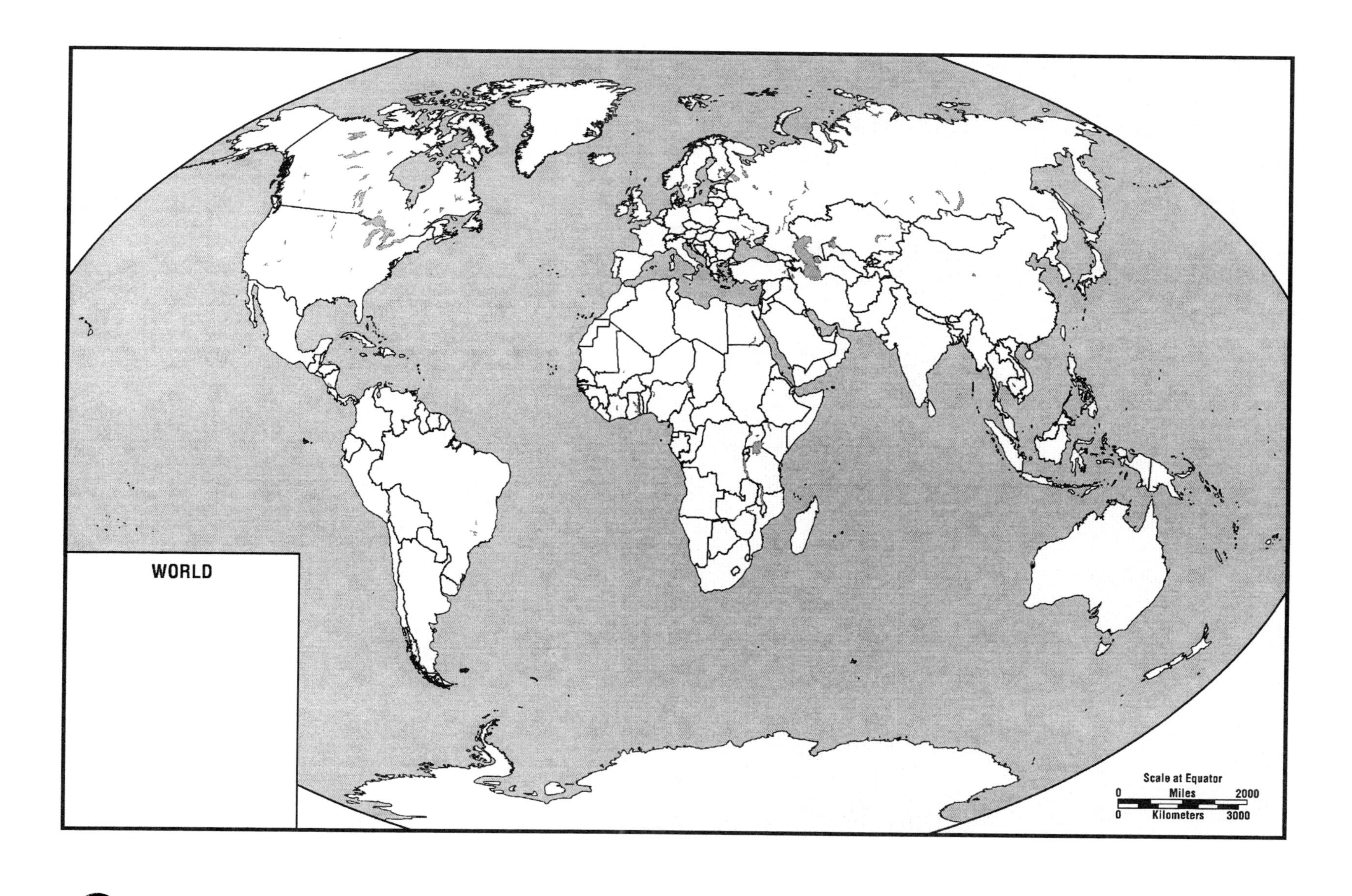
WORLD
Scale at Equator
0 Miles 2000
0 Kilometers 3000